Broadband Access Technology, Interfaces, and Management

For a listing of recent titles in the *Artech House Telecommunications Library,* turn to the back of this book.

Broadband Access Technology, Interfaces, and Management

Alex Gillespie

Artech House
Boston • London
www.artechhouse.com

Library of Congress Cataloging-in-Publication Data
Gillespie, Alex.
　　Broadband access technology, interfaces, and management / Alex Gillespie.
　　　　p.　cm. — (Artech House telecommunications library)
　　Includes bibliographical references and index.
　　ISBN 0-89006-473-3 (alk. paper)
　　　1. Broadband communication systems.　2. Computer network protocols.
　　3. Computer networks—Management.　4. Computer network architectures.
　　I. Title.　II. Series.
　　TK5103.4. G56 2000 2001
　　621.382'1—dc21

00-064279
CIP

British Library Cataloguing in Publication Data
Gillespie, Alex
　　Broadband access technology, interfaces, and management. —
　　(Artech House telecommunications library)
　　1. Broadband communication systems
　　I. Title
　　621.3'981

　　ISBN 0-89006-473-3

Cover and text design by Darrell Judd

© 2001 ARTECH HOUSE, INC.
685 Canton Street
Norwood, MA 02062

International Standard Book Number: 0-89006-473-3
Library of Congress Catalog Card Number: 00-064279

10 9 8 7 6 5 4 3 2 1　　　　**3 3001 00849 1770**

Contents

Contents			**v**

Preface			**xxiii**
	Conventions		*xxiv*
	Acknowledgments		*xxiv*

1	**Overview**		**1**
1.1	Broadband and ATM		2
1.2	The Evolution Toward Broadband		3
1.3	Access Networks, Core Networks, and Service Providers		3
1.4	Broadband Technology		4
1.4.1	Access Network Technology		4
1.4.2	Upgrading the Core Network		7
1.4.3	The Boundary Between Core and Access Networks		8
1.5	Summary		9

2	**Network Architecture**	**11**
2.1	Computer Networks	12
2.1.1	Repeaters, Bridges, and Routers	12
2.1.2	Internets, Intranets, and the Internet	13
2.1.3	Remote Internet Access	14
2.2	Broadband Access and Service Providers	14
2.2.1	Service and Service Provider Interfaces	15
2.2.2	End-User Choice and Network Complexity	16
2.3	Broadband Access Architecture	16
2.3.1	Technology, Services, and Dimensioning	16
2.3.2	Services and the Core Architecture	18
2.4	The ATM Core	18
2.4.1	Permanent Virtual Connections	19
2.4.2	Switched Virtual Connections	20
2.5	Summary	21
	References	22

3	**Economic Considerations**	**23**
3.1	Broadband Service Considerations	23
3.1.1	Video-on-Demand (VoD)	24
3.1.2	Internet Access	25
3.1.1	Video Telephony	27
3.2	Broadband Connection Considerations	28
3.3	Summary	29

4	**SONET/SDH**	**31**
4.1	The Historical Background	31
4.1.1	The Plesiochronous Digital Hierarchies	32
4.1.2	Synchronous Transport Signals	33
4.2	Multiplexing and Architecture	34

4.2.1	The Three Laws	34
4.2.2	From Tributaries to Synchronous Interfaces	36
4.2.3	The Functional Architecture	41
4.3	Adaptation Overheads and Synchronization	43
4.3.1	Synchronization and Coordination	44
4.3.2	Pointers and Rate Adaptation Channels	45
4.4	Layer-Specific Overheads and OAM Flows	46
4.4.1	Common OAM Functions	47
4.4.2	Section Layer Overheads	48
4.4.3	Path Layer Overheads	48
4.5	The Management Interface and Management Model	49
4.5.1	Objects and Relationships in the Management Model	50
4.5.2	Transactions for the Management Model	53
4.6	Summary	54
	References	55
5	**ATM Fundamentals and Management Modeling**	**57**
5.1	Paths, Channels, and Cells	57
5.1.1	Physical Paths, Virtual Paths, and Virtual Channels	58
5.1.2	The Format of ATM Cells	58
5.1.3	Comments on ATM Cells	60
5.2	ATM Layers and Functions	61
5.2.1	The Transmission Convergence Layer	62
5.2.2	The VP and VC Layers	63
5.2.3	Higher Order Layers	63
5.3	The Basic Management Model for ATM	64
5.3.1	ATM Configuration Management	64
5.3.2	ATM Performance Management	68

5.4	Summary	69
	References	71

6	**ATM Operations, Administration, and Maintenance (OAM) Flows**	**73**

6.1	OAM Flow Layers and Ranges	74
6.2	OAM Cell Types and Functions	76
6.2.1	Fault Management	77
6.2.2	Performance Management	79
6.2.3	Activation/Deactivation	82
6.3	The Operations System (OS) Interface	84
6.3.1	The Notification for Defect Reporting	84
6.3.2	Loopback	85
6.3.3	Continuity Checking	85
6.3.4	Performance Monitoring	87
6.4	Problems and Deficiencies	88
6.5	Summary	90
	References	90

7	**ATM Adaptation for Client Services**	**91**

7.1	Introduction	91
7.1.1	The Structure of the ATM Adaptation Layers	92
7.1.2	Management Modeling of ATM Adaptation	93
7.2	Synchronous Traffic (AAL1 and AAL2)	94
7.2.1	AAL1 Segmentation and Reassembly	95
7.2.2	AAL1 Synchronization Modes	96
7.2.3	The AAL1 Convergence Sublayer (CS)	97
7.2.4	AAL1 Management and Modeling	98
7.2.5	Synchronous Variable Rate Traffic (AAL2)	99
7.3	Asynchronous Traffic—Original (AAL3/4)	100
7.3.1	AAL3/4 Segmentation and Reassembly	100

7.3.2	AAL3/4 Convergence Sublayer (CS)	101
7.3.3	AAL3/4 Management and Modeling	102
7.4	Asynchronous Traffic—Streamlined (AAL5)	104
7.4.1	AAL5 Segmentation and Reassembly	104
7.4.2	The Common Part of the AAL5 Convergence Sublayer	105
7.4.3	AAL5 Management and Modeling	106
7.4.4	Internet Protocol (IP) over AAL5	107
7.5	The Signaling ATM Adaptation Layer (SAAL)	107
7.5.1	The Service-Specific Connection-Oriented Protocol (SSCOP)	108
7.5.2	The Service-Specific Coordination Functions (SSCFs) for the SAAL	112
7.5.3	Management Modeling of the SAAL	113
7.6	Comments on ATM Adaptation	114
7.7	Summary	114
	References	115
8	**ATM Signaling**	**117**
8.1	Background	117
8.1.1	ITU-T and the ATM Forum	118
8.1.2	Interworking	119
8.2	Services, Addresses, and Topology	119
8.2.1	Bearer Services and Telecommunications Services (Teleservices)	119
8.2.2	ATM Addresses	120
8.2.3	Connection Topologies	121
8.3	UNI Signaling	122
8.3.1	ITU-T UNI Signaling	122
8.3.2	ATM Forum UNI Signaling	124
8.4	NNI Signaling	124

8.4.1	ITU-T NNI Signaling	125
8.4.2	ATM Forum Intra-NNI Signaling	129
8.4.3	ATM Forum Inter-NNI Signaling	131
8.5	Summary	131
	References	132

9   **Management of ATM Switches**   **133**

9.1	Background	133
9.2	ATM Interfaces	134
9.2.1	Modeling of UNIs	134
9.2.2	Modeling of NNIs	135
9.3	Service Profiles	136
9.3.1	Bearer Services, Teleservices, and Supplementary Services	137
9.3.2	Modeling of Supplementary Services	138
9.3.3	Customizing Services for Addresses	140
9.3.4	The Circuit Emulation Service	140
9.4	Configuration of the Routing Algorithm	141
9.4.1	Modification of Destination Addresses	142
9.4.2	Local Destinations	142
9.4.3	Abstract (Remote) Destinations	143
9.4.4	Routes to Remote Destinations (Post Analysis Evaluation)	143
9.5	Summary	144
	References	145

10   **Internet Communication**   **147**

10.1	Introduction	147
10.2	IP Addresses and Address Resolution	149
10.2.1	The Address Resolution Protocol (ARP)	150
10.2.2	The Reverse Address Resolution Protocol (RARP)	152

10.2.3	Multicasting on the Internet	153
10.2.4	Advanced Address Registration: BOOTP and DHCP	154
10.3	IP Over ATM	155
10.3.1	ATM Link-Level Addresses	156
10.3.2	The ATM Address Resolution Protocol (ATMARP)	157
10.4	Internet Control Messages	159
10.4.1	Control on the Local Network	160
10.4.2	Reporting Discarded Datagrams	161
10.4.3	End-to-End Control	163
10.5	End-to-End Data Transport	164
10.5.1	The User Datagram Protocol (UDP)	165
10.5.2	The Transmission Control Protocol (TCP)	167
10.6	Routing	171
10.6.1	Algorithms: Distance Vector and Shortest Path	171
10.6.2	Autonomous Systems, Gateway Protocols, and the Internet Core	172
10.6.3	The Exterior Gateway Protocol (EGP)	173
10.6.4	The Routing Information Protocol (RIP) and Routing Loop Avoidance	176
10.6.5	The Border Gateway Protocol (BGP)	176
10.6.6	Open Shortest Path First (OSPF)	177
10.7	Summary	177
	References	179
11	**Internet Applications**	**181**
11.1	Introduction	181
11.2	User-Friendly Addresses: The Domain Name System	182
11.3	The Internet Trinity	184

11.3.1	Remote Login: TELNET	184
11.3.2	The File Transfer Protocol (FTP)	185
11.3.3	Electronic Mail (e-mail)	186
11.4	Hypertext and the World Wide Web	187
11.4.1	HTTP Messages	188
11.4.2	HTTP Methods	188
11.4.3	Responses to HTTP Requests	189
11.5	Remote Procedure Calls	190
11.6	Summary	191
	References	193
12	**Management of the Internet (SNMP)**	**195**
12.1	Messages and MIBs	196
12.1.1	Abstract Syntax Notation One (ASN.1) and SNMP Messages	196
12.1.2	MIBs and Internet Management	198
12.2	Basic MIB-II Groups and Their Evolution	198
12.2.1	The System Group	198
12.2.2	The Interfaces Group	198
12.2.3	The Internet Protocol (IP) Group	201
12.2.4	The Internet Control Message Protocol (ICMP) Group	201
12.2.5	The Transmission Control Protocol (TCP) Group	201
12.2.6	The User Datagram Protocol (UDP) Group	202
12.2.7	The Exterior Gateway Protocol (EGP) Group	202
12.2.8	The Simple Network Management Protocol (SNMP) Group	203
12.3	Additional Groups	203
12.3.1	Internet Additions	203
12.3.2	Network Groups	205
12.3.3	Other Groups	206
12.4	The Technology-Specific Groups	207

12.5	Summary	207
	References	208

13   ADSL Transmission   211

13.1	Tones, Modulation, and Coding	212
13.1.1	Tones	212
13.1.2	Modulation	213
13.3.3	Trellis Coding	214
13.2	Frames, Superframes, and Symbols	216
13.2.1	Frames, Superframes, and Tone Allocation	216
13.2.2	Payload Frames and ADSL Channels	216
13.3	Forward Error Correction	217
13.3.1	Error Detection and Error Correction	217
13.3.2	Forward Error Correction in ADSL	219
13.4	Channels, Ports, and Framing	219
13.4.1	Bearer Channels, ADSL Channels, and ATM Ports	219
13.4.2	Framing and Overheads	220
13.5	Summary	221
	References	222

14   ADSL Management   223

14.1	The ADSL MIB Module	224
14.1.1	Technology-Independent Information	224
14.1.2	ADSL-Specific Information	224
14.2	The CMIP Model for the Management of ADSL	229
14.2.1	Configuration, Status, and Alarms	231
14.2.2	Profiles	231
14.2.3	Performance Monitoring	231
14.3	Summary	234
	References	234

15   VB5 Access Architecture   237

15.1   Service Nodes (SNs) and VB5 Interfaces   237

15.2   Logical Ports and Physical Ports   238
15.2.1   Logical User Ports (LUPs)   239
15.2.2   The Logical Service Port (LSP)   240
15.2.3   System Configuration   241

15.3   Signaling and UNI Accesses   242
15.3.1   The VB5 Protocols   242
15.3.2   User Signaling and UNI Accesses   244

15.4   Comments on the VB5 Architecture   244

15.5   Summary   245
         References   245

16   VB5 Protocols   247

16.1   VB5 Messages and Message Format   247

16.2   Protocol Errors   249
16.2.1   The Protocol Error Cause Information Element   250

16.3   The Real-Time Management Coordination
         (RTMC) Protocol   250
16.3.1   Some Problems with I.610   250
16.3.2   RTMC Messages   251
16.3.3   RTMC Information Elements   261

16.4   The Broadband Bearer Connection Control
         (B-BCC) Protocol   263
16.4.1   B-BCC Messages   263
16.4.2   B-BCC Information Elements   271
16.4.3   B-BCC Protocol Anomalies   273

16.5   Summary   275
         References   275

17	**VB5 Management**	**277**
17.1	Background	277
17.2	VPs, Logical User Ports, and Logical Service Ports	279
17.2.1	VP Level Configuration in the Service Node	279
17.2.2	VP Level Configuration in the Access Network	281
17.2.3	Configuration of VB5 Protocols	281
17.3	The Relationship with VB5 Messages	282
17.3.1	Start-Up of an Interface	282
17.3.2	Checking the Interface	283
17.3.3	Resetting the VB5 Interface	283
17.3.4	The State of Resources	283
17.4	Broadband Access Coordination: X-VB5	284
17.4.1	Specific X-VB5 Transactions	284
17.4.2	Generalized X-VB5 Transactions	285
17.4.3	RPC Specification of X-VB5 Transaction Requirements	286
17.5	Summary	287
	References	288
18	**Optical Access**	**289**
18.1	Background	289
18.2	ATM PON Architecture	290
18.2.1	ATM PON Transmission	291
18.2.2	VDSL Transmission on Hybrid Architectures	293
18.3	PLOAM Cells on ATM PONs	294
18.3.1	Common PLOAM Fields	295
18.3.2	Downstream-Only PLOAM Fields	296
18.3.3	Upstream-Only PLOAM Fields	297
18.4	OLT/ONU Coordination	298

| 18.5 | Summary | 298 |
| | References | 299 |

| **19** | **ATM Enhancements** | **301** |

| 19.1 | Background | 301 |

| 19.2 | Enhanced OAM Flows | 302 |

19.3	ATM (Automatic Protection Switching) APS	304
19.3.1	Types of Protection Switching	304
19.3.2	The Protection Protocol	305
19.3.3	Conditions, Commands, and States	306

19.4	Paths and Connections	307
19.4.1	Switched Virtual Paths	307
19.4.2	Soft Permanent Virtual Connections (S-PVCs)	308
19.4.3	Multipoint Connections	309

19.5	Traffic, Services, and Quality	311
19.5.1	ATMF Service Categories and ITU-T Transfer Capabilities	312
19.5.2	Traffic Parameters, QoS Parameters, and QoS Classes	313

| 19.6 | Summary | 314 |
| | References | 315 |

| **20** | **Optical Technology for IP** | **317** |

| 20.1 | Background | 317 |

| 20.2 | IP over Serial Data Links | 318 |

20.3	Optical IP Transmission	320
20.3.1	IP over SONET/SDH	320
20.3.2	ATM Versus SONET/SDH	321
20.3.3	IP over WDM	322

20.4   Optical IP Networks   323

20.4.1   Optical Burst Switching   323

20.4.2   Optical Flow Switching   324

20.4.3   WDM LANs and WANs   325

20.4.4   Superimposed Optical Topologies   326

20.5   Summary   326

         References   327

21   The Way Forward   329

21.1   Background   330

21.2   Multiprotocol Label Switching   330

21.2.1   Forwarding Equivalent Classes   331

21.2.2   Creating Label-Switched Paths (LSPs)   332

21.2.3   MPLS Versus ATM   333

21.3   Internet Protocol Version 6   334

21.3.1   Datagram Format   334

21.3.2   Fields and Headers   334

21.4   The Resource Reservation Protocol   336

21.5   Shortcut Routing   337

21.5.1   Policy-Based Shortcuts   338

21.5.2   Next Hop Resolution Protocol   339

21.5.3   Topology-Independent Shortcuts   340

21.6   RSVP for ATM   341

21.6.1   Quality in IP and ATM   341

21.6.2   Making Shortcuts with Reservations   342

21.6.3   Proxy Addresses and VB5   343

21.7   Summary   343

         References   345

Acronyms and Abbreviations   **347**

About the Author   **363**

Index   **365**

Preface

No man but a blockhead ever wrote, except for money.
Dr. Samuel Johnson

The future growth of the Internet requires greater data rates for remote access than can be achieved with narrowband technology. Not only does broadband access allow faster download of Web pages and e-mails, it also encourages new types of service, such as Video-on-Demand, that do not rely on Internet communication.

The book is concerned with the technology of broadband access, and how this technology is monitored and controlled by operations systems. The book has four parts. It starts with introductory chapters that are followed by an examination of existing ATM, SONET/SDH, and IP technologies. Part Three covers recent developments in ADSL and the VB5 interface, while Part Four looks to the future of optical technology, ATM, and IP, closing with some provocative ideas about future communications protocols.

The division between Internet service providers and other providers with broadband access is not hard and fast. This is because providers of both types of service do not want to limit their options and are prepared to move in or out of Internet-based services. Unlike narrowband access, which can be free, it is likely that no one but a blockhead will ever provide broadband access, except for money.

Conventions

In various chapters of this book, management models for broadband technologies are described. The diagrams that show the relationships between managed objects use certain conventions (see Figure P.1). Different symbols are used for single objects, multiple objects, and optional objects. Contained objects are shown within the objects that contain them, and arrows are used to represent pointers between objects. The thickness of the heads and tails of the arrows indicate if there are single or multiple objects at either end.

Acknowledgments

Many of my colleagues have helped and encouraged me during the writing of this book. Although I would prefer to mention some by name, this would be unfair to many others. There is also no point since it will not encourage them to buy the book as they are willing to wait as long as necessary for me to provide a free copy. However, I would like to express my specific thanks to the reviewer that Artech House assigned, whose comments have been especially useful.

My family has also given me considerable help and encouragement. P.G. Wodehouse once dedicated a book to his daughter, Leonora, for her never-failing sympathy and encouragement without which the book would have been completed in half the time. This book is dedicated more simply to my son, Philip, for being willing to postpone his swimming lessons.

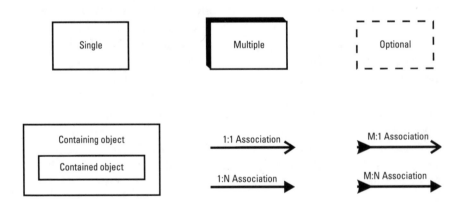

Figure P.1 Conventions in management models.

1

Overview

From the moment I picked it up until I laid it down,
I was convulsed with laughter. Some day I intend reading it.
Groucho Marx (attributed)

Broadband access is the link between service providers and the end-users of their broadband services. It is rather like Groucho Marx, who is supposed to have said that he would not join the sort of club that would accept someone like him as a member. Two clubs that might accept broadband access are the Internet club and the not-narrowband club. Unfortunately one of these is too exclusive and the other is not a proper club at all.

The not-narrowband club is not a proper club because it is the world outside of the narrowband club. Narrowband communications are built from 64 Kbps channels. These channels may be combined for higher capacity services. Although there is no hard and fast bit rate that divides broadband from narrowband, there is a practical difference because narrowband services use narrowband switches. These are either analog switches or their digital equivalents that switch 64 Kpbs connections. Unfortunately, there are services, such as broadcast television, that are neither narrowband nor truly broadband.

The Internet club is the club that everyone wants to belong to. The Internet club would like to have broadband access since it is hungry for bandwidth and would like to guarantee quality of service. The need for bandwidth is especially strong in the infrastructure of the Internet to ensure that its rapid growth does not create bottlenecks. Despite the popularity of the Internet club, it is not the club for everyone as there are some things, like Video-on-Demand (VoD), at which its members do not shine.

The first part of this book is an introduction to broadband access, containing an overview and examining the economics and network architecture. The second part gives more detail about mature technologies, in particular the established aspects of SONET/SDH, Asynchronous Transfer Mode (ATM), and the Internet Protocol (IP).

The third part of the book looks at the more recent developments, in particular ADSL and the VB5 interface. The fourth and final part looks at recent developments in optical access, ATM, and IP, concluding with some provoking ideas for the future.

First we take a step back to get a feel for the wider picture.

1.1 Broadband and ATM

Some people find it difficult to think of broadband as anything other than ATM, partly because of the adoption of ATM for broadband ISDN by the ITU-T. Local Area Network (LAN) technologies demonstrate that broadband is not confined to ATM, but they are not appropriate to broadband access because they are used for local communications within organizations, not for connecting end-users to service providers. The difference here is that between connecting to a corporate Ethernet in the office and remote access to an Internet Service Provider (ISP) from home.

The idea that the Internet might supersede ATM will be examined in the final part of this book. For now it is enough to note that the Internet is not in direct conflict with ATM because they each do different things. The Internet provides communications protocols that are not tied to a network technology. ATM is a transport technology that is not tied to a physical transmission.

It is possible for broadband access to use transmission technologies directly without ATM. This can be done both for optical fiber and for copper pair transmission. Likewise IP can be used directly over SONET/SDH, and over optical and electrical media. ATM is also not necessary when cable modems are used for broadband access.

ATM can still be taken as the basis for broadband access so long as the willingness remains to consider alternatives and to adopt them if they are more sensible.

1.2 The Evolution Toward Broadband

Broadband transmission has been easy to justify in the core of the telecommunications network because of the volume of traffic that the core network handles. Broadband transmission for the access is less easy to justify. This is because there needs to be a demand for high capacity communications by end-users to justify having broadband links to them.

The development of ATM has helped to fuel this demand. It has also reversed the historical trend because ATM is more of a revolution than an evolution. Broadband transmission evolved to handle large numbers of narrowband connections between telephone exchanges. ATM was deployed in islands for stand-alone applications.

Both ATM islands and non-ATM LANs exist on the periphery of the telecommunications network. The initial driver for the deployment of ATM in the core network has been the need to interconnect these islands and LANs.

1.3 Access Networks, Core Networks, and Service Providers

Telecommunications operators often view broadband networks as split into two parts, namely an access network and a core network. The end-users of the broadband services are at the far ends of the access network. The core network often includes the link to the service provider because its capacity can be similar to other core links.

The core network directs broadband traffic between its nodes. The access network carries the broadband traffic between end-users and a node of the core network. Unfortunately the boundary between the access network and the core network is not clear-cut because both the access network and the core nodes have used similar technology.

The core network carries the broadband traffic across large geographical areas. It can be connection oriented, connectionless, or a mixture of the two. If it is connection-oriented then connections can be semi-permanent or can be established and released on-demand. Switch nodes join the transmission links to create the end-to-end connections.

Connectionless networks do not create end-to-end connections. Instead they have routers that direct the flow of individual packets of data. The transmission links between switch node and routers are also part of the core network.

1.4 Broadband Technology

The gray area between core networks and access networks has become better defined as the technologies have developed. Both IP routers and ATM switches that handle on-demand connections belong to the core part of public broadband networks. Advanced transmission technologies that connect end-users to the core network belong to the access network.

1.4.1 Access Network Technology

Phase one of the deployment of broadband technology for access networks is the use of legacy transmission media with new transmission technology. Phase two involves the widespread deployment of new transmission media. Between these two phases there can be the deployment of hybrid architectures that compensate for the shortcomings of the legacy media (see Figure 1.1).

1.4.1.1 Phase One: Legacy Media

The first phase of the development of broadband technology for the access network involves using the same physical transmission medium as an existing service, but with a new transmission technology. Asymmetric Digital Subscriber Line (ADSL) transmission is a good example of such a new technology. ADSL operates over the line that carries ordinary telephony. The other common service that demonstrates this is cable television where the new technology is cable modems.

Although coaxial cable is the better physical medium for broadband transmission, it is not possible to take full advantage of this because it is shared between a large number of users (see Figure 1.2). It is the better physical medium because it has a greater bandwidth and more uniform transmission characteristics. It is shared between a large number of cable television

	PHASE ONE: legacy media	TRANSITION: compensating	PHASE TWO: widespread new
Telecoms	DSL (esp. ADSL)	Fibre-to-the-Cabinet	Hybrid media
Cable TV	Cable modems	Hybrid Fiber/Coax	Hybrid media

Figure 1.1 Phase one (legacy media) to phase two (widespread new media).

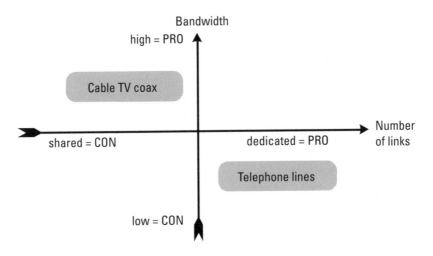

Figure 1.2 Pros and cons of legacy copper media.

users in a tree-and-branch architecture because this is an efficient way to deliver a large number of broadcast television channels.

For telephone lines the situation is reversed. Telephone lines provide a dedicated transmission path to users. The disadvantage of telephone lines is that they have lower bandwidth and more variable transmission characteristics that are suitable for the transmission of analogue voice signals but not good for broadband. It is possible to compensate for the frequency variation of the transmission characteristics by using more sophisticated transmission techniques, but the attenuation at higher frequencies leaves the transmission susceptible to interference.

The cost of a copper pair is much less than the cost of coax because a copper pair is simpler to make and because it takes up less space. The space required is the dominant factor as the dominant cost is that of digging trenches and installing the medium. Copper pairs have an advantage over coax as the same trench can hold many more copper pairs, but this advantage is less than it appears at first because the bandwidth of coax is greater.

1.4.1.2 Transition: Compensating Hybrid Architectures

It is possible to compensate for the limitations of both approaches to the first phase by modifying the architectures to introduce optical fiber at the core end (see Figure 1.3). This has the advantage of supporting the use of the physical transmission media that is already installed, and the disadvantage of the maintenance costs of the opto-electronic equipment in the field.

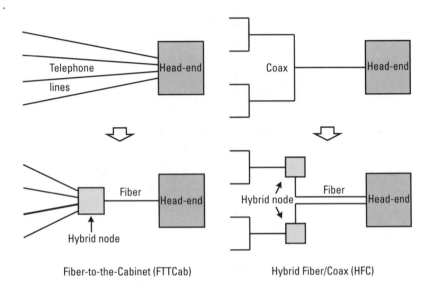

Figure 1.3 Compensating hybrid architectures.

There are also the equipment and cabling costs for the opto-electronic equipment, but in the longer term it is the maintenance costs that dominate the overall costs. The disadvantage of a hybrid architecture is less when optical fiber is used to support coax because coax requires amplifiers in any case to compensate for the attenuation of the cable and the signal reduction when the cable is split.

In a Hybrid Fiber/Coax (HFC) architecture, the optical fiber is used to reduce the number of users that are served by each coaxial tree-and-branch and to provide an independent feed to the coaxial part of the architecture. This compensates for the limitation produced by the users sharing the coaxial bandwidth because it reduces the number of users in each cluster.

Fiber can also be used to compensate for the attenuation of the copper pairs in a Fiber-to-the-Cabinet (FTTCab) architecture by providing the link to a cabinet that is nearer to the user. The transmission between the active node and the user is still over existing telephone lines, but the link between the exchange site and the active node has low attenuation because the physical medium is optical fiber.

1.4.1.3 Phase Two: Widespread New Media

The second phase of technological development for access networks differs from the first phase because it is not limited by the capabilities of the physical media installed for earlier services. The key difference of opinion about the second phase is whether or not it is feasible to include previously installed physical media.

The difference of opinion regarding the role of copper pairs is manifest in the two architectures developed by the Full Services Access Network (FSAN) collaboration. One view is that the second phase should be a variation of the hybrid architecture of the first phase, with the existing copper pairs used for Very-High-Speed Digital Subscriber Line (VDSL) transmission instead of ADSL transmission. The other view is that optical fiber should be deployed all the way to the user.

Unfortunately, the cost of installing a new, fully optical access network for broadband access is significant and has to be justified by the services that it will carry.

1.4.2 Upgrading the Core Network

The core network differs from the access network because it already contains mature broadband technology in the form of ATM cross-connects, SDH transmission, and IP routers. In the short term, the rollout of broadband access only requires more of this technology to be deployed. In the longer term this technology needs to be upgraded.

A key upgrade for ATM (see Figure 1.4) is the introduction of Switched Virtual Connections (SVCs). In an ATM cross-connect, Permanent Virtual Connections (PVCs) are configured by a management system. This is not efficient because capacity is allocated to connections even when it is not needed. A better approach is for ATM connections to be established and released as required using call control signaling. Traditionally call control signaling originates with an end-user. ATM also supports the intermediate step of Soft Permanent Virtual Connections (SPVCs), where the connection appears as a configured connection to the end-users, but call control signaling is used to establish and maintain the connection within the ATM network.

This traditional approach allows an end-user to establish a connection with a broadband service provider when a particular service is required, instead of being permanently connected. For broadband access it is also possible for a service provider to control connections on behalf of an end-user.

Upgrading an ATM cross-connect with configured connections to a full ATM switch that handles call control signaling means adding functionality. Additional functionality is required to allow an ATM switch to use signaling protocols and to interpret the signaling messages according to the

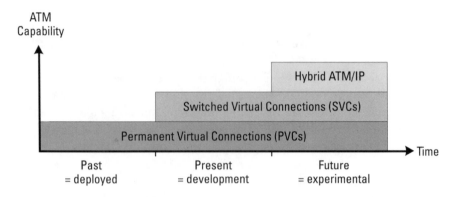

Figure 1.4 ATM developments.

services allocated to the user and the trunk routes that exist within the network.

ATM switches also give greater flexibility to IP routers, since they allow capacity to be allocated to transmission paths as it is needed. This may be essential for the development of the Internet because it can eliminate the bottlenecks that occur when transmission paths become congested since existing paths can be upgraded and new paths created as needed. In the longer term, proposals to integrate ATM switching with IP signaling and routing may be adopted.

1.4.3   The Boundary Between Core and Access Networks

Mature ATM technology in the form of Permanent Virtual Circuits (PVCs) has applications both in the access network and in the core network. In the core network, SDH transmission can be used to carry PVCs between ATM switches and IP routers. PVCs can also be carried over legacy PDH transmission equipment and even directly over new Wavelength Division Multiplexing (WDM) technology without depending on a multiplexing hierarchy. SDH is often preferred because it can provide automatic protection switching that re-routes the SDH path if a link fails, but the introduction of direct ATM protection switching could change this.

In the access network, SDH transmission can be used in feeder networks that connect the multiplexers of optical or electrical transmission systems to the core network. This allows access transmission systems to be located remotely from nodes in the core network. Part of the difficulty in defining the boundary between access and core networks is deciding whether this SDH transmission system belongs to the access network or to the core network. For network operators, defining this boundary can be important because different operating divisions are often involved. The definition is also important to regulatory bodies because it can mark the boundary between different network operators.

Another part of the difficulty in defining the boundary involves the use of ATM cross-connects. For the first generation ATM core network the ATM cross-connects are the means by which the connections are routed. In a second generation ATM core network with Switched Virtual Connections (SVCs) that are established and released on-demand, ATM cross-connects are an effective way of providing flexibility in the configuration of trunk routes.

The access network also performs ATM cross-connection. When this is done within the multiplexer of an access network transmission system it is clearly part of the access network. The problem arises if an ATM cross-connect is used to provide flexibility between the multiplexer and a node within the core network. As in the SDH case, both the access network and the core network can claim ownership, and the boundary can be important for both network operators and regulators.

1.5 Summary

Broadband access is the link between service providers and the end-users of broadband services. Asynchronous Transfer Mode (ATM) is widely accepted as appropriate for this link because ATM is decoupled from the physical transmission technology. ATM was initially deployed in unconnected islands and the first task of the public broadband network was to interconnect these islands and to interconnect non-ATM LANs.

Broadband access is normally regarded as split into two parts, namely the access network and the core network, although there is a gray area that can be regarded as either. The distinction can be important both to network regulators and internally within network operators. The core network is concerned with routing both connectionless and connection-oriented communications across a large geographical area, and can include the links to the service providers. The access network is concerned with the transmission that links end-users to the core network.

Part of the gray area involves the use of mature broadband technologies, in particular ATM cross-connects and SDH transmission that play a key role in the first generation core network. The second generation core network involves upgrading to on-demand ATM connections that make more efficient use of resources and that can assist in removing the bottlenecks from the Internet. In the longer term proposals to integrate ATM and IP routers may be adopted.

Phase one of broadband access in the access network involves using electrical transmission over the copper pairs or coaxial cable used for telephony or cable television. In each case the limitations of the existing architectures can be reduced through hybrid optical/electrical transmission systems. In phase two, the restrictions of the legacy media are eliminated through a greater use of optical fiber.

In the remainder of this book there is a fuller introduction to broadband access that is followed by examinations of mature broadband technology, recent developments, and provocative ideas for the future. The emphasis is less on the cable modems introduced by cable operators than on the evolution that this has helped to stimulate in existing narrowband access to Internet Service Providers (ISPs).

2

Network Architecture

There are three rules for writing a novel. Unfortunately, no one knows what they are.

—W. Somerset Maugham

The 1999 prospectus of the Department of Architecture at Cambridge University in England suggests that the study of architecture is also suitable for those who intend a career in other fields, such as the media. Unfortunately, the architectural design of the network for broadband access appears to offer even more scope for those of a creative nature.

Luckily there are some rules that can be applied. Broadband access is access to something and it is access for a reason. Specifically, it is access to some other network for services provided on that network. If this is a computer network then part of the architecture must be consistent with the architecture of computer networks. Since the end-users of the service are remote, there must be a transmission technology on the links to the end-users. There must also be a broadband communications network that links the end-users with the provider of the service.

These are the rules, and they are good rules. They can be used to guide the architectural design for broadband access. Unfortunately, no one knows for sure that they are the right rules.

2.1   Computer Networks

Connectionless communications networks can be classified as Local Area Networks (LANs) or Wide Area Networks (WANs). Typically a LAN consists of a passive transmission medium that computers connect to through a network interface card that handles the LAN protocol. The area covered by a LAN is about the size of a university campus or a business site, but over this area it can offer high data rates and low delays. Ethernet is a good example of a LAN technology.

A WAN allows communications over a larger distance than a LAN. A WAN contains routers that are connected by communications links. The range of a WAN is not limited by the physical constraints of the transmission medium, but its delays are greater and its data rates are less than those of a LAN. The data rates are less because they are limited by the cost of the capacity on the communication links. The delays are greater because of the delays in the communications links, and these are especially large if satellite links are used.

2.1.1   Repeaters, Bridges, and Routers

The range of a LAN can be extended through repeaters and bridges, and LANs in different locations can be linked through routers (see Figure 2.1).

A router is a computer that routes packets between different networks. It is physically connected to different networks and knows how they carry packets. It also knows, from the destination address of a packet, which network to route the packet to. The job of a router is much easier if the packets have the same format, such as the format of an Internet datagram.

A bridge is a computer that passes frames between the physical segments of the same LAN. The size of LAN segments is limited by the physical properties of the medium and of the signals transmitted on it. A bridge recovers frames and passes valid frames to the other side. Since a bridge regenerates the frames it recovers, it restores degraded signals to their original form.

Most bridges also decide which frames to pass. If this is done intelligently by an algorithm, the bridge is called an adaptive bridge. When an adaptive bridge receives a frame, it notes its source address and the side that sent it. It then knows that it does not have to pass a frame for this destination to the other side of the bridge.

Bridges need additional intelligence to handle broadcast frames in a complex topology. This is necessary to avoid a chain reaction when several

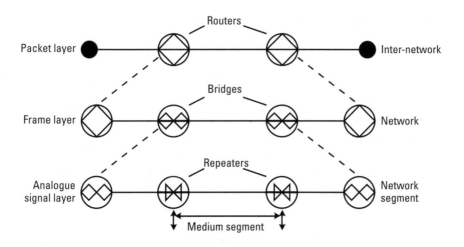

Figure 2.1 Repeaters, bridges, and routers.

bridges deliver the same frame to a segment. Without coordination, each bridge may return the copies that other bridges have delivered and the other bridges may send these back to the segment later. With suitable co-ordination, bridges are effective means of extending a LAN, especially if most of the traffic is local to each segment.

A repeater is not a computer, but a hardware device. Repeaters operate at the analogue signal layer to relay signals from one physical segment to another. Repeaters amplify the signals they receive and they may compensate for distortion. Repeaters do not eliminate noise, errors, or invalid frames as bridges do because they do not recover the digital information. The signals that they transmit may also collide with signals on the destination segment.

2.1.2   Internets, Intranets, and the Internet

An internet is simply a linked group of networks. An intranet is a linked group of networks that belongs to the same organization. Normally information that can be accessed within an intranet cannot be accessed from the outside.

The Internet is the global group of linked networks that use the Internet Protocol (IP). An internet or intranet need not use IP, and need not be connected to the Internet. An intranet that is linked to the Internet normally has a firewall of constraints that limit access.

2.1.3   Remote Internet Access

The most common form of remote Internet access is a telephone connection. This is a point-to-point connection that uses modems to carry a stream of digital information. This architecture is quite different from the architecture of an Ethernet LAN.

Point-to-point connections between end-users and a router can be treated either as multiple networks or as a single network. If they are treated as multiple networks then the only machines on each network are the end-user machine and the router. If they are treated as a single network then they share a single Internet network address, but the router has to determine which physical interface to use when a datagram arrives for an end-user.

It is more common for each connection to be treated as a separate network. The Serial Line Internet Protocol (SLIP) or the Point-to-Point Protocol (PPP) is used to frame the Internet datagrams [1]. These protocols can also be used for broadband access to the Internet since they are independent of the transmission rate.

For broadband access from a PC an ADSL modem can be used (see Figure 2.2). If this modem is not integrated into the PC then it is sensible for it to have a universal serial bus (USB) interface because this is faster than a serial or parallel interface. An ATM layer [2] above the physical transport avoids the need for routers between the different physical access links.

2.2   Broadband Access and Service Providers

Narrowband access to an Internet Service Provider (ISP) is via a PSTN or narrowband ISDN connection. This connection may go all the way to the

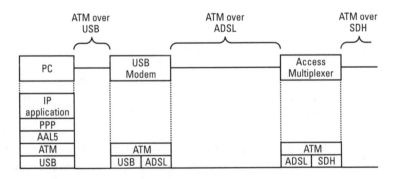

Figure 2.2   Remote broadband Internet access.

ISP (see Figure 2.3). Alternatively the narrowband connection may terminate in a Remote Access Service (RAS) that belongs to the telecommunications network operator with Internet communication to the ISP.

2.2.1 Service and Service Provider Interfaces

Broadband access requires higher data rates to end-users than narrowband. This can be achieved over existing telephone lines using Asymmetrical Digital Subscriber Line (ADSL) transmission (see Chapter 13). The development of ADSL followed on from that of Digital Subscriber Line (DSL) technology for narrowband ISDN. Unlike ISDN, ADSL does not include call control and uses permanent connections to a multiplexer. Even higher rates can be achieved with optical access technology, either all of the way to the end-user or with VDSL transmission on the final copper drop (see Chapter 20). Broadband access using radio transmission is also possible, but this is less common because radio bandwidth is a limited resource.

Broadband access is not restricted to IP communications because the higher data rates also directly support services such as Video-on-Demand (VoD). In this case the interface to the Service Provider (SP) is at the transport layer, and this is likely to be an ATM layer so that it is decoupled from the physical transport (see Figure 2.4).

IP can also be supported directly by the physical link to the end-user so long AS it provides framing for IP datagrams. This means that a SP interface can be at any layer from the physical layer upwards, depending on its location in the network. Telecommunications operators are keen to have interfaces at the higher layers that provide added value, whereas regulators and the IP community are more interested in interfaces at the lower layers that give a greater break from traditional telecommunications.

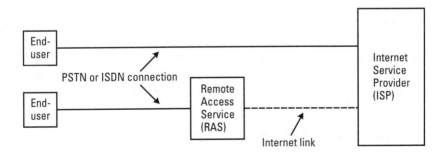

Figure 2.3 Options for narrowband access to an Internet Service Provider (ISP).

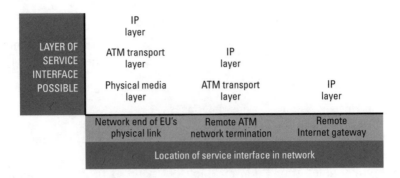

Figure 2.4 Layers and locations of Service Provider (SP) interfaces.

2.2.2 End-User Choice and Network Complexity

The relationship between end-users and SPs is a key consideration in the design of the network. The design is simpler if end-users have a permanent connection to a SP. The more choice that is given to end-users, the more complex the network must be.

An end-user that is permanently connected to a single SP may be offered a choice of services. This requires a greater range of communications because different services, such as VoD and Internet access, have different connection requirements. Alternatively, the end-user may be able to choose a SP from a fixed group (see Figure 2.5). For example, an end-user requiring Internet access could have a number of permanent ATM connections, each to an ISP, delivered over an ADSL link and be able to select an ISP simply by using the correct connection.

The network becomes even more complex if an end-user has an unrestricted choice of both services and SPs.

2.3 Broadband Access Architecture

It is easier to define the access architecture than to dimension it. This is because the architecture is determined by the technology, while the dimensioning is determined by the service. This is illustrated by the case of ADSL.

2.3.1 Technology, Services, and Dimensioning

ADSL implies a star architecture because ADSL operates over telephone lines. These lines feed a multiplexer that connects to a core broadband

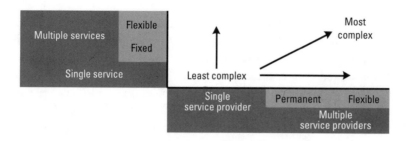

Figure 2.5 Services, service providers, and network complexity.

network (see Figure 2.6). It is natural to ask about the size of the multiplexer and the rate of the link to the core.

The link to the core poses an immediate problem because telecommunication links are normally symmetrical. ADSL transmission achieves its range by restricting the upstream transmission rate to reduce the limit imposed by crosstalk. If there is a symmetrical link between the access multiplexer and the core network then much of its upstream capacity will be unused. As we shall see, much of the downstream capacity may also be unused.

SDH is the obvious choice for physical transmission to the core. Comparing the minimum SDH rate to the ADSL rate should give the minimum size of the multiplexer. The nature of the services and of ADSL prevents this.

ADSL transmission need not operate at fixed rates. ADSL can be configured for a variety of rates in either direction and it can also adapt to find the highest rate. For simplicity, upstream and downstream rates may be artificially fixed for all lines.

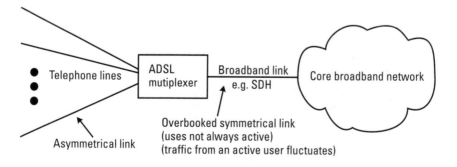

Figure 2.6 Broadband access architecture for ADSL.

The service need not operate at the full ADSL line rate and different services have different requirements. A VoD service requires a constant high downstream rate. Surfing the Internet is more bursty because pages are downloaded and then examined. It is not efficient to reserve bandwidth for all end-users at all times because at any time only some end-users will require service. Those who do will often generate fluctuating traffic. This means that the link to the core network can support many more end-users (see Figure 2.6).

The downstream link to the access multiplexer is likely to be under-utilized because of mismatch between the potential of the link and the physical size of the multiplexer.

2.3.2   Services and the Core Architecture

If all of the traffic from each access multiplexer is to a single SP and the SDH links with the multiplexers are well utilized then there is no need for an ATM core network because it is more sensible to have direct SDH links to the SPs.

ATM is useful as the core network technology if the links to the ADSL multiplexers are under-utilized because it can consolidate traffic from several multiplexers onto the link to the SP. Likewise if a single multiplexer supports several SPs then an ATM core can distribute the traffic to them.

2.4   The ATM Core

An ATM core improves the efficiency of the network by allowing traffic from access multiplexers to be groomed. It also provides an infrastructure that is decoupled from the physical transport and can carry a wide variety of traffic.

An ATM connection is either a Permanent Virtual Connection (PVC) or a Switched Virtual Connection (SVC). A PVC exists for the duration of the contract regardless of whether or not there is traffic present. An SVC is created as and when it is needed.

At first sight, an SVC core seems obviously superior to a PVC one since switched connections are the proven approach for telephony. The situation here is different from conventional telephony because most connections are between end-users and SPs.

The simplest ATM connection between and end-user and a SP is one that operates at the Virtual Path (VP) layer. This architecture is not efficient because many of these connections follow the same routes through the network (see Figure 2.7) and only differ on the final leg to the end-users. It is more efficient to consolidate the end-user connections onto a shared VPC.

2.4.1 Permanent Virtual Connections

A PVC architecture is more efficient than it initially appears because there is no need to interconnect every node. This is because the number of SPs is small, and not all end-users are connected to every SP.

A PVC architecture can also be overbooked so that bandwidth on a PVC is not dedicated to end-users. This is more effective if permanent Virtual Channel Connections (VCCs) are used in addition to permanent Virtual Path Connections (VPCs) because the VPCs can be used to consolidate and groom traffic from the VCCs.

2.4.1.1 Permanent Virtual Path Connections (P-VPCs)

It is difficult to overbook physical links with P-VPCs since the physical links do not span the ATM network. Overbooking of the end-user link is feasible if the end-user traffic is with different SPs at different times.

Overbooking of links in the core network is less easy because this requires knowledge of the service traffic on the links, and this is difficult unless each connection between an end-user and aN SP is tagged with a service identifier. Overbooking of the core connections is easier to achieve if the connections between end-users and the SPs ARE at the VC level, at least at the periphery of the core network.

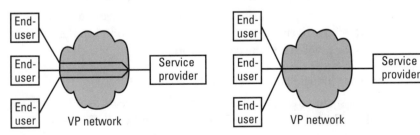

Figure 2.7 Separate and shared VP connection to service provider.

2.4.1.2   Permanent Virtual Channel Connections (P-VCCs)

P-VCCs between a SP and the end-users on the same access multiplexer can be carried across the core network on the same P-VPC. These can be over-booked with P-VCCs according to the nature of the service. It is easier to calculate the overbooking for each shared P-VPC because each can be designed to carry the same type of service.

The grooming of VC traffic onto service specific P-VPCs is achieved through VC cross-connections in the access multiplexers. This allows an existing ATM core network that only supports VP level connections to be used efficiently without the introduction of VC level cross-connection in the core.

2.4.2   Switched Virtual Connections

SVCs also raise the architectural issue of the relative levels of functionality in the access multiplexer and in the core network, particularly with respect to call control signaling.

2.4.2.1   Switched Virtual Channel Connections (S-VCCs)

If the access multiplexer does no more than cross-connection then the call control signaling for each end-user must be processed in the core network. This is far from ideal because bandwidth on the links between the multiplexer and the core is not used efficiently and because the core network must be upgraded to handle signaling for all end-users. The VB5.2 version of the VB5 interface (see Chapter 15) improves the efficiency of the links, but the core network must still be able to handle all of the user signaling.

There is a case for handling user signaling in the access multiplexer because this reduces the level of upgrade needed in the core network. In this architecture, the access network becomes an access switch with functionality similar to that of a PABX. This access switch can have a single User-Network Interface (UNI) into the core network with all end-users treated as sub-users on this UNI. Alternatively there can be a Network-Node Interface (NNI) between the access switch and the core network.

2.4.2.2   Switched Virtual Path Connections (S-VPCs)

S-VPCs have a slight advantage over S-VCCs because fewer connections are needed at the VP level since each S-VPC carries several VCCs. This advantage is not great because often only a small number of VCCs are needed for each service, and because the bandwidth of the S-VPC may need to be modified as VCCs within it are used.

S-VPCs seems to combine the worse features of both P-VPCs and S-VCCs. The simplicity of overbooked P-VPCs has disappeared because signaling been introduced and the precision in the allocation of bandwidth is less than for S-VCCs.

2.5 Summary

The ranges of the local area networks (LANs) that link computers are limited by their technologies. The range of a LAN can be extended by using repeaters, bridges, and routers. Wide area networks (WANs) contain routers that are joined by communications links. Remote Internet access has a different architecture from a LAN because it uses individual point-to-point connection.

Traditionally, remote computers have used telephone lines and modems to access a computer network or Internet Service Provider (ISP) from a remote location. ADSL technology and other new access technologies allow remote access at broadband rates. The more choice that is given to the end-users, the more complex the network architecture needs to be.

The transmission technology largely determines the end-user architecture, but efficient dimensioning of the architecture is difficult because of the variety of end-user services. This is aggravated when end-users are connected using ADSL transmission because the traditional technologies for the network interface of the access multiplexer are symmetrical. A core ATM network can be useful because it can consolidate traffic from different access multiplexers onto the same physical link to an SP and groom traffic from one multiplexer onto several SPs.

ATM connections can be permanently configured or switched according to demand. The conventional wisdom that switched architectures are more efficient than permanent connection between all nodes is undermined if end-users are only connected to SPs. Overbooking of capacity increases the efficiency of a permanent virtual connection (PVC) architecture especially if the virtual path (VP) layer is overbooked with virtual channel (VC) layer traffic and is used to groom traffic. Overbooking of the end-user link with VP traffic is also possible if an end-user is unlikely to need access to all SPs simultaneously.

The level of functionality of the access multiplexer is an even greater issue for switched virtual connection (SVC) architectures than for PVC architectures because an access multiplexer could process user signaling. Less functionality is needed if the access multiplexer has a VB5 interface, but in this case the ATM core still needs to process the user signal. Switching at the end-user connections at the VP layer rather than at the VC layer does not significantly reduce the number of connections and prevents the grooming of VC traffic into VPs.

References

[1]   Tanenbaum, A. S., *Computer Networks*, Upper Saddle River, NJ: Prentice-Hall, 1996.

[2]   McDysan, D. E., and D. L. Spohn, *ATM Theory and Application*, New York: McGraw-Hill, 1995.

3

Economic Considerations

Marketing managers can't change a lightbulb. They live in perpetual darkness.
— Anonymous

How many customers does it take to justify broadband access? The answer to this depends on whether the customers are service providers or end users. A better question is "How much revenue does it take to justify broadband access?" This time the answer depends on the technology since it is easier to justify broadband access if an existing infrastructure can be re-used.

But there is no point in deploying technology for its own sake. Before we can justify broadband access, we need to understand the business case. The emphasis in the first two questions is misplaced. The proper question is "What profit will be generated by broadband access?"

Broadband access provides the link between an end user and a broadband service provider, so to answer the question, we need to examine the services that are supported by broadband access. It is the revenue that is generated by these services that determines whether broadband delivery is viable.

3.1 Broadband Service Considerations

The history of digital services has not been devoid of hype. To be fair, some of the more imaginative ideas are not fundamentally unsound, but have required advances in technology that were not feasible in the time-scales being considered. However, in the early days of the narrowband Integrated Services Digital Network (ISDN) it was suggested that it would be used for video telephony. Subsequent advances in video compression techniques have made this feasible and High-speed Digital Subscriber Line (HDSL) transmission allows it to have good quality. Now after two decades of ISDN there is no widespread use of video telephony.

The reason for the slow take-up of video telephony is what people feel about it. This illustrates the importance of looking carefully at a service before deciding that we need the technology.

3.1.1   Video-on-Demand (VoD)

Broadband access must support Video-on-Demand (VoD), but VoD does not justify the development of broadband access because the revenues are too small. A ballpark estimate of these revenues can be obtained from the charge for renting a video from a shop and the number of videos that can be watched over a period of time.

A VoD service can charge more for a video than a shop charges for rental because VoD is more convenient and can offer a larger choice of videos with more recent releases. Even so, the charge per video for a VoD service is unlikely to be much more that twice that for shop rental because people will use the alternatives if the charge is higher. These alternatives include going to a cinema, ordering the video from a library, waiting until it is available for shop rental, and buying it outright.

The number of videos that can be watched in a period of time is also limited. People already record films that are broadcast on television and find that they do not have time to watch them later. Allowing for other forms of entertainment, perhaps two or three videos a week might be supplied from a VoD service to each private residence. In the long term the average could be less because the VoD library may become exhausted once people had watched their favorite videos. People who want to watch a video again and again would probably buy it outright.

The estimated charges and frequency of use of a VoD service suggest that the revenues generated would be no more than those generated by the public switched telephony network (PSTN) service. This figure is confirmed if VoD is compared with satellite, cable or licensed television, since these contend for similar markets. There is additional confirmation from staggered broadcasting from satellites where switching channels is equivalent to fast forwarding or reversing on a VoD service.

VoD may be included in a package of services supported by broadband access because of the additional revenue that it can generate. There is also competitive pressure to include VoD because comparable services, such as staggered broadcasting and pay-per-view television, are supported by existing networks whose operators wish to expand into other areas supported by broadband access.

The conclusion is that the business case for developing broadband access based on a VoD service alone is weak unless broadband access is possible at a cost comparable to PSTN or satellite or cable television.

3.1.2 Internet Access

The Internet has taken the world by storm, and it is now difficult to imagine going back to life without it. Since the mid-1990s, there have been concerns about the demand for the Internet outstripping its capacity. These concerns do not appear to have materialized. This may be because of people's expectations of the Internet, and their expectations of the cost of being on-line.

People expect that electronic mail will sometimes go missing, and may even rely on this. If an e-mail is important then a second e-mail will be sent, or there will be a follow-up phone call or fax. If there is no follow-up then the e-mail cannot have been important, and it does not really matter whether it has been lost or simply ignored or deleted by mistake. If e-mail is successful most of the time then there is a significant benefit and it is possible to work around the occasional failure.

There are similar expectations about access to Web sites. It can be inconvenient if the access times are slow, but again this is something that it is possible to work around. If it is important to a business to have rapid access to Web pages then the business can add capacity to its own intranet and mirror external sources of information. If it is important to an individual to access Web pages then it can be scheduled to times of day when access is faster. Again there is a significant net benefit that does not rely on the Internet being perfect.

These expectations are reflected in the prices people are willing to pay for Internet access. Typically there is either a one-off or a monthly charge from an Internet Service Provider (ISP). A monthly charge may be fixed or it may depend on usage. In any case there is normally a connection charge that depends on the duration of the call to the ISP. Although the total cost of narrowband Internet access varies from user to user, the ballpark figure for the total is similar to that for PSTN service.

For the residential market, the acceptable costs of broadband Internet access are of the same order of magnitude as for narrowband Internet access. This is because of narrowband ISDN, Internet bottlenecks, consumer perceptions of technological evolutions, and the cost of CD-ROMs.

The introduction of D-channel access for narrowband ISDN creates an upper limit for the cost of narrowband Internet access because it reduces or eliminates the charges that depend on call duration since the cost of introducing usage-based charging for packets on the D-channel is prohibitive. It is easier to include the charge for using an ISDN D-channel for the connection to an ISP in the ISDN line rental. There is still a component that remains dependent on the duration of the connection if ISDN B-channels are brought up to increase the data rate or if the ISP charges for usage. In any case, the cost will be less than for modem access over the PSTN network because the connection time will be less. The result is good news for heavy users of narrowband Internet access, and bad news for the business case for broadband Internet access.

There is more bad news for broadband Internet access because of the bottlenecks in the Internet core. There is little advantage to having a high capacity link to an ISP if the data rate across the Internet to the desired Web page is poor. In the longer term this will not be a problem because the capacity in the Internet core should increase, but this is little consolation to a user who receives no immediate perceptible benefit for a significant increase in the cost of access. This does not affect the business market because there are service guarantees for LAN interconnect and virtual private networks (VPNs), but these are reflected in costs that are not acceptable to residential users.

A significant increase in cost is unlikely to be high on a residential user's agenda since it is contrary to the way in which technological enhancements are perceived. When a new technology is first introduced, people may be prepared to pay a high price for it because it does something that was not previously practical. From then on, prices are expected to fall as the market grows because of improvements in technology and economies of scale. Technological enhancements, in contrast to technological improvements, are a way to maintain the initial high prices. Users would expect the cost of broadband Internet access to be of the same order as the initial cost of narrowband Internet access because they see it as an enhancement to an existing service and not a new service.

CD-ROMs allow a sanity check on this view because of their speed of access and because of the availability of writable CD-ROMs. A writable CD-ROM can be used to mirror information from Web sites using narrowband Internet access. CD-ROMs can also be used as sources of information comparable to Internet Web sites. In both cases the information stored locally on a CD-ROM can be accessed at a rate at least as great broadband access to a remote Web site. People are unlikely to pay an order of magnitude more for broadband Internet access than they would for the same amount of information on a CD-ROM, since all they are paying for is convenience.

This all suggests that in the residential market people would only be prepared to pay a percentage premium for broadband Internet access above the cost of narrowband access. The step from narrowband Internet access to broadband access is more like the step from black-and-white television to color television than from radio to black-and-white television. The future residential market differs from the existing business market for LAN interconnects where connections are charged according to bandwidth.

3.1.1 Video Telephony

In contrast to Internet access, the step from voice telephony to video telephony is superficially similar to the step from radio to television. This similarity is superficial because people were eager to see television broadcasts whereas people are reluctant to use video telephony.

There is a barrier between people using voice telephony that makes them comfortable with it. During a voice call it is possible to do things that during a video call would be disruptive or discourteous. A video call is much more like being in the same room. In particular, activities that involve a loss of eye contact, such as checking the time, looking out of a window, or turning to gesture to someone else, are much more disruptive on a video call than on a voice call.

A video call is also more intrusive than a voice call. This is because it is more like having someone in your own territory, someone who is able to gauge your reactions and expressions. The invasion of privacy in a video call is especially unwelcome in residential applications because it allows strangers and intruders to see the condition and layout of your home. In some cases, adults would even be reluctant to allow increased visibility of their living conditions to their parents, and would be uncomfortable about explaining the reason why.

Video telephony makes videoconferences easier, but this is unlikely to drive the development of broadband access because it is a low volume, specialized application. There are also doubts about the advantages of videoconferences over conventional meetings because they do not provide the same opportunities for socializing and having off-line discussions. Ironically, the problem here is the opposite of that with voice telephony since video telephony offers too much interaction in comparison with voice telephony, but too little interaction in comparison with conventional meetings.

Video telephony as a driver for broadband access is also questionable because video telephony can also be delivered over narrowband ISDN. The quality of video telephony over narrowband ISDN is not as good as that over a broadband connection, but it is adequate for video telephony since the main requirement is for a slowly changing picture of a head and shoulders. The option of narrowband video telephony makes video telephony similar to Internet access because broadband video telephony is perceived as an enhancement of narrowband video telephony, and not a new service.

3.2   Broadband Connection Considerations

There is downward pressure on the call charges for broadband video telephony not only because narrowband ISDN supports video telephony, but also because of the intrinsic logic of broadband connections. Broadband connections cannot be charged in proportion to their data rate with respect to narrowband connections because this would make them too expensive. Alternatively, if broadband connections were charged at the rate that residential market would support, then charging by data rate would mean that narrowband charges would have to be slashed to an uneconomic level.

The way out of this is to argue that the dominating cost of any connection is not the data rate, but the call processing cost. Whether this is true is not the issue here, the point is that the argument allows operators to justify maintaining their charges for narrowband connections while introducing broadband connections with a much higher data rate at a much lower charge per bit. Unfortunately for operators, this argument also reinforces the view that broadband is an enhancement of narrowband since the charges for both are essentially the call processing costs.

This view in turn reinforces the view that the broadband network cannot effectively replace the narrowband network. It is not only the decades of developments in the narrowband network that would have to be duplicated, but the broadband network itself is not well suited to handling narrowband corrections. The reason for this is that the broadband network introduces additional transmission delays for narrowband connections due to the time required to assemble Asynchronous Transfer Mode (ATM) cells from the lower rate narrowband sources. This is because the cells are too large to handle narrowband connections well. The outcome is that additional echo-cancellation functionality needs to be added to each broadband connection that is used to carry narrowband voice traffic.

3.3 Summary

Despite the hype, it is not possible to identify any single service as a commercial driver for residential broadband access that will generate revenues that are an order of magnitude greater than those of narrowband services. Competitive pressure from similar services limits the revenues from Video-on-Demand (VoD). The charges for narrowband Internet access limit the revenue from broadband Internet access. The low demand and the availability of narrowband video telephony limit the revenues from broadband video telephony. In addition, the call charges for broadband connections will have to be less per bit than for narrowband connections.

For many years, perhaps decades, the broadband network will exist in parallel with the narrowband network because broadband is not optimized for voice communication. However in the residential market, it will not be possible to charge more than a percentage premium over narrowband service changes because there is no sufficiently high revenue service that is specific to broadband.

There are consolations. Residential broadband access can generate significant revenue, comparable to that from existing narrowband services. The diversity of the services it supports makes broadband access less susceptible to misjudgments about an individual service. Services can be demonstrated to be viable with narrowband access before they are upgraded to broadband access. Furthermore there is an established business market for LAN interconnect that can take advantage of new technologies developed for residential broadband access.

4

SONET/SDH

Don't Panic.
—Douglas Adams, *The Hitchhiker's Guide to the Galaxy*

Perhaps the greatest mystery about SONET/SDH is why it is complicated. Its complexity is puzzling because devising and implementing a multiplexing structure for a synchronous network is simple enough to be an undergraduate project.

Half of the reason for the complexity is that the structure of SONET/SDH is determined less by synchronous multiplexing than by its ability to carry traffic from unsynchronized tributaries. The other reason for the complexity is that SONET/SDH was given a strongly hierarchical structure with multiple functional layers.

Faced with this complexity, the first step towards understanding is not to panic.

4.1 The Historical Background

There is both fact and fallacy in the accepted wisdom on why SDH was needed. The fact is that there was a need for a common global multiplexing hierarchy to bridge the historical differences between different regional approaches. These historical differences made international links more difficult than was necessary and made it impossible for equipment suppliers to export their regional products.

The fallacy is more curious and quaint. It was said that the existing plesiochronous multiplexing was complex to implement because it was necessary to demultiplex the entire hierarchy to access a single tributary since there was independent rate adaptation for each layer in the hierarchy. This is true, but there is the wonderful hidden assumption that SDH would be simpler to implement. In addition, the Plesiochronous Digital Hierarchy (PDH) was already well understood and its costs were rapidly declining with very large scale integration of its functionality onto silicon.

Now if I were an equipment supplier then I would want something new and fancy to make more money on, wouldn't you? In reality, of course, it is fortunate that suppliers take a more enlightened and altruistic view than us private individuals.

4.1.1 The Plesiochronous Digital Hierarchies

From the view of the plesiochronous hierarchies, all regions have a common approach at level zero of the hierarchy because level zero is always a 64 Kbps channel. This is only part of the story because there are also regional differences in the way that speech is digitized into 64 Kbps timeslots.

Speech is sampled every 125 μs because the frequency of the highest sound needed for intelligible speech is less than 4 kHz. If these samples were digitized linearly then 16 bits would needed to give the necessary resolution, but linear digitization is not efficient because the human ear is not sensitive to slight variations in the volume of sounds when the volume is high. This means that if an approximately logarithmic scale is used then only 8 bits are needed to represent speech samples. Two logarithmic scales are defined, μ-law and A-law, and different regions use different laws for digitization.

The differences in the digitization laws at level zero of the plesiochronous hierarchies are matched by differences in the data rates at level one. The 1,544 Kbps DS1 rate, used in North America and Japan, consists of 24 timeslots using μ-law digitization (see Figure 4.1). The 2,048 Kbps E1 rate, used in Europe, consists of 32 timeslots and 30 of these may contain speech using A-law digitization. The different hierarchies also use different approaches for signaling and for frame synchronization. Japan also uses DS1 rate, but Japan is a special case since it has its own approach at levels three and four.

For level two, both hierarchies use a four-fold multiplex of their level one rates and again use different approaches for signaling and frame

Rate	Name	Structure	VC	TU/AU
64 kBit/s	DS0 / E0	1 voice channel	n/a	n/a
1544 kBit/s	DS1	24xDS0	VC-11	TU-11
2048 kBit/s	E1	32xDS0	VC-12	TU-12
6312 kBit/s	DS2	4xDS1	VC-2	TU-2
8448 kBit/s	E2	4xE1	n/a	n/a
34,368 MBit/s	E3	4xE2	VC-3	AU-3
44,736 MBit/s	DS3	7xDS2	VC-3	TU-3/AU-3
139,264 MBit/s	E4	3xDS3/4xE3	VC-4	AU-4
	AUG	3xAU-3/1xAU-4		
51.84 MBit/s	STS-1	1xTUG-3		
155.52 MBit/s	STS-3/STM-1	3xSTS-1/1xAUG		
622.08 MBit/s	STS-12/STM-4	4x (STS-3/STM-1)		
2488.32 MBit/s	STS-36/STM-16	16x (STS-3/STM-1)		

Figure 4.1   Multiplexing rates and structures.

synchronization. The North American/Japanese level two rate is 6,312 Kbps and the European level two rate is 8,448 Kbps.

For level three, the different hierarchies use different degrees of multiplexing. The North American/Japanese approach uses a seven-fold multiplex of the level two rate, to give 44.736 Mbps (DS3). The European approach uses a second four-fold multiplex, to give 34.368 Mbps (E3).

At level three, the trend at level one and two for the European rates to be higher than the North American/Japanese rates is reversed. At level four the process goes full cycle and both hierarchies revert to using the same rate again as they did initially at level zero, but now the rate is 139.264 Mbps. This corresponds to a three-fold multiplex of the North American/Japanese level three rate and a four-fold multiplex of the European level three rate.

4.1.2   Synchronous Transport Signals

The Synchronous Optical Network (SONET) was the forerunner of SDH [1]. It was originally developed as an ANSI (American National Standards Institute) standard and operates at 51.84 Mbps with a payload of either a single level three plesiochronous stream or 28 level one plesiochronous streams.

The original structure for this is known as Synchronous Transport Signal level one (STS-1).

The original structure for SONET was altered during the development of SDH. Since the revised structure of SONET corresponds to one configuration of SDH the term SONET is preferred in North America.

4.2   Multiplexing and Architecture

The SONET/SDH multiplexing structure [2, 3] allows different data rates to be carried in a common transmission system [4], and the structure is reflected in the functional architecture [5] and the management model.

4.2.1   The Three Laws

The multiplexing structure can be described by three laws and one clarification. The clarification is that the synchronous nature of SONET/SDH refers to its transmission, not its payloads. The multiplexing structure could be much simpler if the traffic was synchronized to the transmission.

4.2.1.1   The First Law

The first law governs overheads and pointers: Payloads are made into virtual containers (VCs) by adding a path overhead, then low-level VCs are made into tributary units (TUs) and high-level VCs made into administrative units (AUs) by adding a pointer.

VCs are the means by which SONET/SDH carries payloads (see Figure 4.2). They allow path overheads that are key to the management of the multiplexing layers to be added. The overheads carry alarms, performance information, and auxiliary management communications channels.

The TUs and AUs allow payloads at the nominal rates that are not synchronized with SONET/SDH transmission to be supported because they include a buffer zone and a pointer that allow the payload to slip with respect to the SONET/SDH frames. The buffer allows extra data to be sent when the payload frames are running fast.

4.2.1.2   The Second Law

The second law governs low-level payloads (see Figure 4.3), in particular the 1,544 Kbps and 2,048 Kbps tributaries: TUs are made from low-level payloads and are grouped into Tributary Unit Groups (TUGs) that are payloads

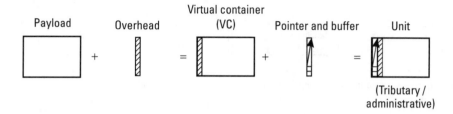

Figure 4.2 Payloads, containers, and units.

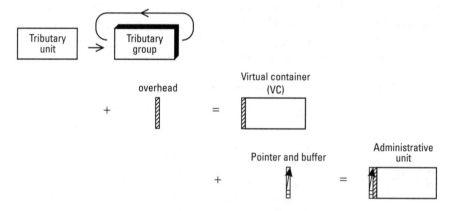

Figure 4.3 Tributary units to administrative units (AUs).

of the high-level VCs that are made into AUs. VC-3 is an anomaly because it can be carried in either a TU-3 or in an AU-3.

4.2.1.3 The Third Law

The third law governs the formation of interfaces: A SONET/SDH interface is a Synchronous Transport Module (STM) consisting of Administrative Unit Groups (AUGs) plus section overheads.

Additional overheads must be added to AUs before physical transmission to carry physical layer alarms and performance monitoring. The SONET/SDH interfaces are known as STM-Ns (see Figure 4.4) where N indicates the number of groups of AUs on the interface. AUs are grouped so that higher rate interfaces can be constructed by simple synchronous multiplexing of the lower rate interfaces.

Figure 4.4   Administrative unit(s) to synchronous transport module(s).

4.2.2   From Tributaries to Synchronous Interfaces

The three laws govern the way that low-level tributary payloads are multiplexed onto synchronous interfaces. The tributary payloads are made into low-level VCs and then into TUs and TUGs. The TUGs are then made into AUs and AUGs, which are made into the SONET/SDH interfaces.

The process starts with the plesiochronous tributaries.

4.2.2.1   Tributaries to Low-Level Virtual Containers (VCs)

According to the first law, tributaries are made into VCs by adding the appropriate management overhead. The three true low-level VCs can be made in this way, as can the hybrid VC-3 container.

The lowest rate tributaries are the DS1 and E1 level one plesiochronous rates. Overheads are added to these using a 4-frame multiframe because this reduces the bandwidth needed for the management overhead. Two octets per frame are added to the 24 timeslots of the DS1 streams to create a VC-11 with 26 octets per frame. Three octets per frame are added to the 32 timeslots of the E1 streams to create a VC-12 with 35 octets per frame (see Figure 4.5).

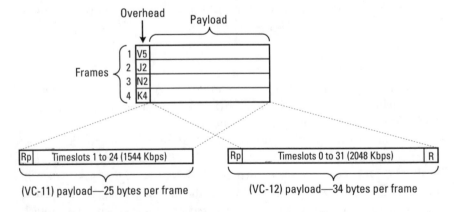

Figure 4.5   Payloads and overheads for VC-11 and VC-12.

It is also possible to use a VC-12 to carry a DS1 stream. This is the preferred approach when the dominant payloads are E1 streams because it can be more effective to avoid VC-11 VCs despite the resulting bandwidth inefficiency since the result is more consistent and has a finer granularity.

The DS2 streams at 6,312 Kbps are equivalent to four DS1 streams at 1,544 Kbps. Eleven octets per frame are added to the 96 timeslots of a DS2 streams to create a VC-2 with 107 octets per frame.

4.2.2.2   Lower-Order Virtual Containers to Tributary Unit Groups

The sizes of VC-11, VC-12 and VC-2 were chosen so that all three can be carried in TUG-2s. This becomes clearer when the pointer is added to the VCs to create TUs in accordance with the first law.

For each of these three VCs, a single octet per frame is added for the pointer and its associated buffer. This gives 27 octets per frame for TU-11s, 36 octets per frame for TU-12s, and 108 octets per frame for TU-2s (see Figure 4.6).

The lowest rate TU is TUG-2 because there is no lower common tributary rate that could be used for both DS1 at 1,544 Kbps and E1 at 2,048 Kbps. Each TUG-2 frame contains 108 octets. This means that each TUG-2 frame takes exactly four TU-11s, or three TU-12s, or one TU-2 (see Figures 4.7 and 4.8).

The payloads of VC-3s and VC-4s are either the level three and four streams of the plesiochronous hierarchies or TUGs that carry a number of level one or two streams. This means that there is flexibility in the way that level one and two plesiochronous streams are carried since they can either be carried as VC-11s, VC-12s or VC-2s or multiplexed into level three and four plesiochronous streams.

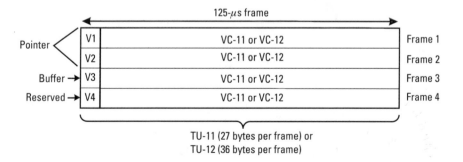

Figure 4.6   Pointers and buffers in TU-11 and TU-12.

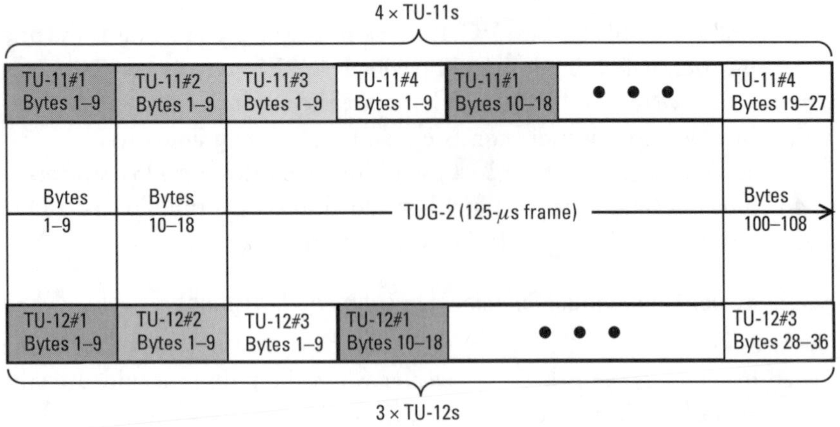

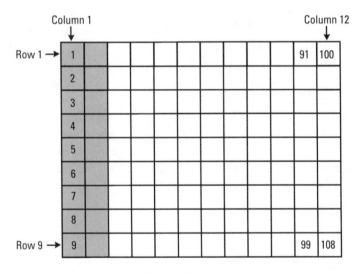

Figure 4.7   From tributary units (TU-11s or TU-12s) to tributary groups (TUG-2s).

Figure 4.8   Bytes into rows and columns for TUG-2.

The flexibility in the way that plesiochronous streams are carried is useful because VC-2 is only suitable for the level two stream for North America/Japan. SONET/SDH can only carry the European level two plesiochronous stream after it is multiplexed onto a higher level of the plesiochronous hierarchy because the 8,448 Kbps European rate is being phased out.

A VC-3 can carry either a single level three plesiochronous stream or seven TUG-2s (see Figure 4.9). The factor of seven here matches the sevenfold multiplexing from level two to level three of the North American / Japanese plesiochronous hierarchy.

Each VC-3 can act either as a lower-order VC or as a higher-order VC. The hybrid nature of VC-3 is deliberate because it allows the SONET configuration of SDH to be used in North America / Japan and the European Telecommunications Standards Institute (ETSI) configuration of SDH to be used in Europe. In Europe, VC-3 is a lower-order VC for carrying level three plesiochronous streams. For SONET, VC-3 is a higher-order VC that is made into an AU-3 in accordance with the second law. If VC-3 is acting as a lower-order VC then it is made into a TU-3 in accordance with the first law (see Figure 4.10).

A TUG-3 consists of either a single TU-3 or seven TU-2s (see Figure 4.11). A VC-4 can carry either three TUG-3s or a single level four plesiochronous stream. The factor of three here matches the three-fold multiplexing from level three to level four of the North American/Japanese plesiochronous hierarchy.

4.2.2.4 Administrative Units (AUs) and Administrative Unit Groups (AUGs)

AUs are made from higher-order VCs by adding a pointer. For the SONET configuration AU-3s are made from VC-3s. AU-4s are only made from VC-4s (see Figure 4.12).

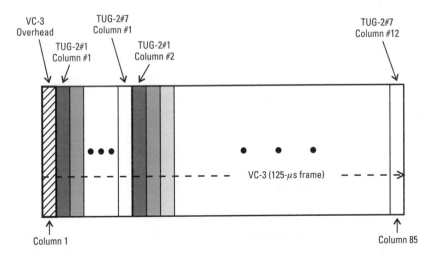

Figure 4.9 From tributary groups (TUG-2s) to virtual containers (VC-3s).

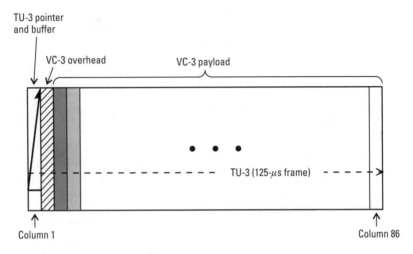

Figure 4.10 From virtual containers (VC-3s) to tributary units (TU-3s).

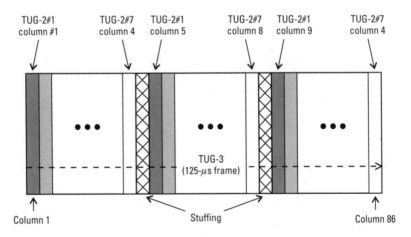

Figure 4.11 From TUG-2s TUG-3s.

An AUG is made from three AU-3s or a single AU-4.

4.2.2.1 SONET/SDH Interfaces

The third law states that each sychronous interface is an STM and the level of the STM indicates the number of AUGs in the interface. The level of the module is a power of four, from one to 64.

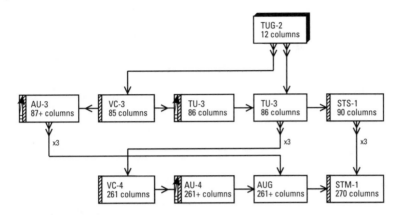

Figure 4.12 Between TUG-2s and STM-1s.

There is also a special case, STM-0, which only applies to SONET configurations. This is one third of the rate of an STM-1 interface, allowing it to carry a single AU-3. The STM-0 interface is the revised form of the SONET STS-1 interface.

The SONET/SDH interfaces are made by adding the regenerator and multiplex section overheads to the AUG. These overheads help to define the section layers of the functional architecture.

4.2.3 The Functional Architecture

The functional architecture of consists of a sequence of layers (see Figure 4.13). On each layer there are trails that consist of a series of connections. Connections are the clients that are served by the trails of the next layer down in the sequence.

Trail termination functions are performed at the end-points of the trails. Adaptation functions are performed between the client and the server layers to adapt the connections in the client layer to the trails in the server layer. There are points between the adaptation functions between two layers and the trail termination functions of both the client and the server layers. Those points on the client layer are called connection points, and those on the server layer are called access points.

4.2.3.1 Architectural Layers

The transport layers serve the circuit layers that contains the plesiochronous connections that are the tributaries for multiplexing. The transport layers

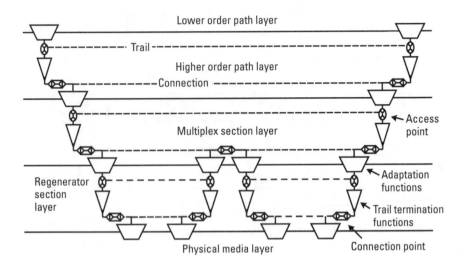

Figure 4.13 SONET/SDH layers and functions.

consist of the path layers and the transmission media layers, with the path layers being the clients that are served by the transmission media layers.

The path layers in turn serve the circuit layers, and they are divided into lower-order path layers and higher-order path layers. This division of the path layers corresponds to separation of the lower and higher order VCs because the trails of the path layers correspond to the transport of VCs. The lower-order path layers are served by the higher-order path layers and are not served directly by the transmission media layers. The lower order path layers can only serve the circuit layers. The higher-order path layers can either serve the circuit layers or the lower order path layers.

The transmission media layers consist of the two section layers, the multiplex section layer and the regenerator section layer, and the physical media layer that serves them. The trails on the physical media layer depend on the physical medium used for transmission, e.g. optical fiber, coaxial cable, or radio.

4.2.3.2 Termination and Adaptation Functions

The trail termination functions of the physical media layer perform conversion between signals and logic levels, loss of signal detection, and laser failure for optical transmission. The adaptation functions between the physical media layer and the regenerator section layer include scrambling, descrambling and frame alignment functions.

The assembly and disassembly of groups into units is an adaptation function that is performed between the appropriate layers. The conversion between units and VCs, and the associated pointer handling, is also performed by these adaptation functions.

The adaptation between the multiplex section layer and the higher-order paths layers handles the AU pointers and allows the higher-level VCs to be extracted from the AUs and AUGs. The adaptation between the higher-order and lower-order path layers handles the TU pointers and allows the lower-order VCs to be extracted from the TUs and TUGs.

The handling of layer overheads and the extraction of payloads are trail termination functions. The overheads of the lower-level VCs and the extraction of their plesiochronous payloads are handled by the trail termination functions of the lower-order path level. The overheads of the higher-level VCs and the extraction of either higher-level plesiochronous payloads or of TUG payloads are handled by the trail termination functions of the higher-level path level.

The extraction of the administrative group payloads is handled by the trail termination functions of the multiplex section layer. The overheads of the multiplex section layer and of the regenerator section layer are handled by the corresponding trail termination functions. These are the section overheads that are added to AUGs to turn them into SONET/SDH interfaces.

4.3 Adaptation Overheads and Synchronization

Framing patterns are overheads that are handled by the adaptation functions that interconnect the layers of the functional architecture. The adaptation functions also handle information about the payload.

The rate adaptation pointers are a combination of pointer and rate adaptation channel. They are associated with overheads but, unlike framing patterns and client or payload specific information, they are not considered as part of an overhead associated with a particular layer because they are purely required for adaptation between layers. The justification for not taking the same view of framing patterns is more a matter of history rather than logic.

4.3.1   Synchronization and Coordination

SONET/SDH frame synchronization involves the generation and recovery of framing patterns for the 125 μs frames used universally for synchronous voice communications. This has to be performed for the various types of STM interface.

SONET/SDH multiframe synchronization is also required in the handling of the overheads for VC-11s, VC-12s, and VC-2s. A similar mechanism is also available for use when the payload of the higher-order VCs is ATM cells.

4.3.1.1   Frame Synchronization

Framing bytes are added as part of the regenerator section overhead for all SONET/SDH interfaces. Two bytes, A1 and A2, are used and these are repeated in regenerator section overheads of STM-1 and higher rate interfaces. The transition between the two bytes provides the reference point for frame synchronization of the interfaces.

In the initial form of the interface another byte, C1, was also included in the regenerator section overhead to label the interleaved STMs of an STM-N interface. This was included initially because it allowed independent searching for the A1-A2 transition on each STM as once the transition was identified, the C1 byte then gave the number of the STM. The C1 byte is not necessary because the STMs are identifiable without it since the first STM is aligned with the first of the repeated A2 bytes and so was replaced by the more useful layer specific trace byte, J0, in the subsequent form of the interface.

4.3.1.2   Multiframe Synchronization

The overhead of the higher-order VC-3s and VC-4s has the H4 byte that is used for multiframe synchronization of the lower-order VC-11s, VC-12s, and VC-2s. The H4 byte defines the four-frame multiframe that is used by the overhead and the rate adaptation pointer for the lower-order VCs. The overhead and the rate adaptation pointer, which is a combination of pointer and rate adaptation channel, both have four bytes and both are transmitted as one byte per frame.

The H4 byte is also used to indicate the offset to the start of the first ATM cell when the ATM cells are carried as the payload of the higher-order VC. The H4 byte is not in fact needed for this because the ATM cells have headers that allow cell to be delineated without additional assistance. In this case the information in the H4 byte is redundant and its use in this way creates an opportunity for inconsistency.

4.3.1.3 Coordination of Adaptation Functions

The C2 byte in the higher-order overhead in VC-3s and VC-4s carries information about the payload. This can be used to co-ordinate adaptation functions between at the ends of the higher-order path layer.

4.3.2 Pointers and Rate Adaptation Channels

Rate adaptation is performed in the adaptation between the lower-order path layers and the higher-order path layer, and between the higher-order path layer and the multiplex section layer. It is achieved using a virtual channel pointer and a rate adaptation channel.

The virtual channel pointer indicates the location of the start of the virtual channel in the client layer of the adaptation process. If the data rate of the client virtual channel is higher than that nominally allocated in the server layer, then additional data is sent and the pointer changes. The additional data is sent in the rate adaptation channel and the pointer changes to indicate that the start of virtual channel in the server layer has moved forwards. This is known as positive rate adaptation since more data than usual is sent.

The data rate in the client virtual channel can also be lower than the nominally allocated rate in the server layer. In this case less data is sent and the pointer changes to indicate that the start of virtual channel in the server layer has moved backwards. When less data is sent the data location in the server layer that immediately follows the rate adaptation channel is skipped. This is known as negative rate adaptation because less data than usual is sent.

In theory there is no need for both positive and negative rate adaptation because allocated capacity in the server layer could be arranged to be more or less than that needed by the client layer. The need for the additional rate adaptation channel would be eliminated if the server layer always had additional capacity, and the peak to peak jitter due to lack of synchronization could be reduced. However, it is not sensible to produce spare capacity by reducing the 125 s frame size since this is a universal period for voice communications and a larger peak-to-peak jitter allows a greater range of interworking between networks.

4.3.2.1   Higher-Order Virtual Containers

The bytes of the higher-order rate adaptation pointer are transmitted along with the regenerator and multiplex section overheads and likewise can be extracted once the framing pattern in the regenerator section overhead is detected. These bytes are known as the H1 to H3 bytes.

The H1 and H2 bytes form the pointer that indicated the start of the virtual channel and the H3 bytes form the rate adaptation channel. The number of H1 and H2 bytes in a frame is determined by the number of AU-3s and AU-4s in the STM-N interface. There is one H1 byte and one H2 byte for each AU since each AU has an associated pointer to indicate the start of the VC that it carries.

The number of H3 bytes in a frame only depends of the order of the STM interface and not on whether AU-3s or AU-4s are present. This is because there is one H3 byte for each AU-3 and three H3 bytes for each AU-4. The reason for this is that is allows the capacity of the rate adaptation channel to scale with the capacity of the AU. The capacity of the rate adaptation channel is 64 Kbps for AU-3s and 192 Kbps for AU-4s.

4.3.2.2   Lower-Order Virtual Containers

The four bytes of the lower-order rate adaptation pointer are transmitted one byte per frame in the four-frame multiframe defined by the H4 byte of the higher-order virtual channels. These four bytes are known as the V1 to V4 bytes. The V1 to V3 bytes behave in a similar way to the H1 to H3 bytes of the higher-order rate adaptation pointer and the remaining byte, V4, is reserved for future use.

There is one pointer, consisting of the V1 and V2 bytes, and one rate adaptation channel of 16 kBit/s, consisting of the V3 byte, for each VC-11, VC-12, and VC-2. The main difference from the higher-order case is that there is no variation now in the number of pointers or in the capacity of the rate adaptation channel.

4.4   Layer-Specific Overheads and OAM Flows

Overheads are defined for all of the SONET/SDH specific layers of the functional architecture, that is for the lower and higher order path layers and for the multiples and regenerator section layers. These overheads can carry both auxiliary communications channels and information flows for operations, administration, and maintenance (OAM).

Although pointers and frame synchronization can be thought of as overheads, they are treated separately because they are used in the adaptation between layers and are not part of a layer-specific overhead.

4.4.1 Common OAM Functions

These common OAM functions may be performed by the overheads of different layers, but not all OAM functions are performed by all layers.

4.4.1.1 Fault Indication

A Far End Receiver Fail (FERF) signal, also known as a Remote Defect Indication (RDI) signal, can be returned when a problem is detected on incoming data. FERF or RDI is a backwards fault indication since it is sent backwards to give prior warning of the presence of the fault. Problems that can cause this include a failure on the data, data mismatch, and detection of an Alarm Indication Signal (AIS).

An AIS may be generated on on-going data due to problems detected on incoming data. AIS is a forwards fault indications because it is sent onwards to indicates that the trail has a fault.

4.4.1.2 Performance Monitoring

The performance monitoring function often uses a monitoring byte to perform Bit Interleaved Parity (BIP) checking that treats the data to be monitored as eight interleaved bit streams (i.e. BIP-8). The monitoring byte contains the eight parity bits for the eight streams, and this can be checked against the eight parity bits calculated for the data when it is received.

4.4.1.3 Performance Reporting

The results of the performance monitoring of received data can be reported back to the transmitting end. This is known as performance reporting.

4.4.1.4 Trail Tracing

The trail tracing function allows trails on a layer to be traced and checked. The identity of the access point at the source of the trail is inserted as a source trail termination function and this identity is recovered for checking as a sink trail termination function. The preferred format for this is a 16 byte pattern consisting of a single byte framing flag and a 15 byte E.164 number.

4.4.2   Section Layer Overheads

The section overheads are added to the administrative section group to make an STM interface. The same basic structure is used both for all STMs including STM-0. To achieve this much of the bandwidth available for the section overheads is often unused because the bandwidth supported by STM-0 is the lowest common bandwidth available for all STMs.

The only differences between the section overheads for the various STMs occur in the multiplex section layer overhead for performance reporting and fault indication.

4.4.2.1   The Regenerator Section Overhead

The regenerator section overhead contains three auxiliary communications channels, but it supports only one OAM function. There is a 64 Kbps engineering order wire channel for voice communications between sites, a 64 Kbps user channel that can be used for proprietary alarms, and a 192 Kbps data communications channel that can support a TMN management interface.

The single OAM function supported by the regenerator section overhead is performance monitoring using one byte per frame for BIP-8, and there is no associated performance reporting.

4.4.2.2   The Multiplex Section Overhead

The multiplex section overhead contains two auxiliary communications channels. There is no user channel in the multiplex section overhead, but there is another 64 Kbps engineering order wire for voice communications and the bandwidth of the data communications channel in the multiplex section overhead has been increased to 576 Kbps.

The multiplex section overhead supports the common OAM functions of fault indication, performance monitoring, and trail tracing. It also supports the specialized functions of protection switching coordination, synchronization class indication, and a specific RDI (for the UNI on STM-1 interfaces only).

4.4.3   Path Layer Overheads

The structure of the overheads is different for higher-order path layers and lower-order path layers, although many of the functions are similar. There are nine bytes per frame for VC-3s and VC-4s and four bytes per four-frame multiframe for the lower-order VCs. VC-3s are treated as higher-order as regards their overheads.

4.4.3.1 The Higher-Order Path Overhead

The higher-order path overhead contains a single auxiliary communications channel, a user channel at 64 Kbps that can be used for proprietary signaling. This is an inversion of the situation for the multiplex section overhead where the user channel was the only auxiliary channel present in the regenerator section overhead that was not present in the multiplex section overhead.

The higher-order path overhead also supports the common OAM functions of backward fault indication, performance monitoring, performance reporting, trail tracing, and payload labeling. It also supports specialized functions for protection co-ordination and for performance monitoring of tandem connections

4.4.3.2 The Lower-Order Path Overhead

The lower-order path overhead supports the same the common OAM functions as the higher-order path layer, namely backwards fault indication, performance monitoring, performance reporting, tail tracing and payload labeling. Likewise it also supports specialized OAM functions for protection co-ordination and for the performance monitoring of tandem connections.

Unlike the higher-order path overhead, it does not support any auxiliary communications channels.

4.5 The Management Interface and Management Model

Although other approaches to the management of SONET/SDH have been devised and implemented, in particular the SNMP approach [6], the ITU-T approach uses the Common Management Information Protocol (CMIP) and a specified management model [7]. The structure of this management model is derived from the generic management information model defined in ITU-T Recommendation M.3100 [8]. This generic structure is based on the layers and termination and adaptation functions of a generic functional architecture. The specific structure of the management model is based on the functional architecture.

The relationships between the objects of the management model are represented in one of two ways. Objects can have pointer attributes that identify other objects, or they can be contained within other objects much as directories on a computer can be contained within other directories. Objects can only be directly contained within one other object, but several objects can be contained within the same object and all objects except root objects are contained within other objects in a containment hierarchy. Containment typically indicates that there is an existence dependency in what is being modeled because contained objects cannot exist when the containing objects does not exist. The are no corresponding constraints when relationships are represented by pointers.

4.5.1 Objects and Relationships in the Management Model

The fundamental classes of the management model are associated with the trail termination functions of the layers of the functional architecture and the adaptation functions between these layers. There are also additional classes that are used to model the interfaces for PDH tributaries on SONET/SDH equipment, the cross-connections within SONET/SDH equipment, automatic protection switching for SONET/SDH, and the boards, racks and other physical equipment that supports SONET/SDH functionality.

The root object of the management model directly contains the objects that model the trail termination functions of the various layers, the trail termination point objects (see Figure 4.14). The objects that model the adaptation functions between layers, the connection termination point objects, are contained within the trail termination point objects of the server layer for the adaptation. This is because the adaptation functions (such as frame alignment and pointer handling) that they represent cannot exist unless there are terminated trails. Pointers between the client trail termination point object and the connection termination point object represent the relationship between the trail termination functions of the client layer and the adaptation functions between the client and server layers.

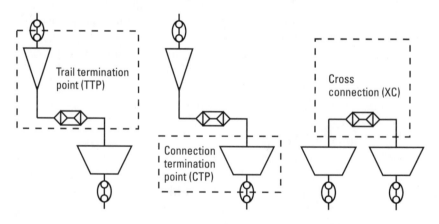

Figure 4.14 Mapping of the functional architecture onto managed objects.

Typically there are separate object classes to represent sink and source termination and adaptation functions. These are also combined to form bidirectional classes that represent combined sink and source functions.

4.5.1.1 Physical Medial Layer to Regenerator Section Layer

The optical and electrical synchronous physical interface trail termination point objects model the trail termination functions of the optical and electrical physical media layers. There is one object for either the transmitter or the receiver or for the combined transmitter and receiver of each physical STM-N interface. These objects are contained within the root object for the SONET/SDH network element.

The regenerator section connection termination point objects model the adaptation functions between the physical media layer and the regenerator section layer. These objects are contained within the corresponding trail termination point objects for the physical interface and point to the client trail termination point objects for the regenerator section layer.

4.5.1.2 Regenerator Section Layer to Multiplex Section Layer

The regenerator section trail termination point objects model the trail termination functions for the regenerator section layer. These objects are contained within the root object for the network element and they point back to the regenerator section connection termination point objects that model the adaptation functions to the physical media layer.

Each trail termination point object for the regenerator section can contain connection termination point objects that model the adaptation to auxiliary layers for the order wire channel, the user channel and the data communications channel. These are the three auxiliary channels carried in the regenerator section overhead.

Trail termination point objects for the regenerator section also contain multiplex section connection termination point objects that model the adaptation functions between the regenerator and the multiplex section layers. These connection point objects point to the client trail termination point objects for the multiplex section layer.

4.5.1.3 Multiplex Section Layer to the Higher-Order Path Layer

The multiplex section trail termination point objects model the trail termination functions for the multiplex section layer. These objects are contained within the root object for the network element and they point back to the multiplex section connection termination point objects that model the adaptation functions to the regenerator section layer.

Each trail termination point object for the multiplex section layer can contain connection termination point objects that model the adaptation to auxiliary layers for the order wire and data communications channels. These are the two auxiliary channels carried in the multiplex section overhead.

Trail termination point objects for the multiplex section also contain AUG objects that model the AUGs of the SONET/SDH interface. These in turn contain AU connection termination point objects that point to the client trail termination point objects for the higher-order path layer.

4.5.1.4   Higher and Lower Order Path Layers

The higher-order VC trail termination point objects model the trail termination functions for the higher-order path layer. These objects are contained within the root object for the network element and they point back to the AU connection termination point objects that model the adaptation functions to the multiplex section layer.

Each trail termination point object for the higher-order path layer can contain connection termination point objects that model the adaptation to the auxiliary layers for the user channel that is carried in the higher-order path overhead. They can also contain level three TUG objects that can in turn contain level two TUG objects or a level three TU connection termination point. A TUG-2 object can in turn contain TU-11 or TU-12 connection termination point objects or a TU-2 connection termination point object.

The TU connection termination points model the adaptation functions to the corresponding path layers. They also point to the corresponding client trail termination point objects that are contained within the root object for the network element and that point back to them.

4.5.1.5   Path Layers to Plesiochronous Tributaries

All VC trail termination point objects can contain the corresponding level of plesiochronous connection termination point objects. These connection termination point objects model the adaptation functions to the plesiochronous layer. This modeling configuration is used when the payload of the VC is the appropriate layer one, two, three or four plesiochronous tributary.

4.5.2 Transactions for the Management Model

The basic CMIP transactions allow objects to be created and deleted, objects to send notifications, the attributes of objects to be read and often written to, and requests to be sent to objects to request the performance of specific actions. In the management model, the main use of action requests is to control the relationships when SONET/SDH cross-connection is being performed.

4.5.2.1 Notifications

Many objects in the management model can send common notifications to indicate when they are created or deleted, when their operational or administrative states change or when their other attributes change. Certain objects may also send a generic alarm notification.

The detection of the loss of framing by the adaptation function typically causes a communication alarm notification to be issued by the connection termination point object that models the adaptation function. This is the generic notification but it contains a probable cause parameter that has a value defined to indicate a loss of frame. Other problems detected by adaptation functions can also result in this notification being sent.

The notification is also sent by trail termination point objects when trail termination functions detect faults. These faults can be detected directly, for example if there is a loss of signal, or indirectly, for example when faults are reported in the overhead.

4.5.2.2 Reading and Writing Attributes

Objects typically have a read-only attribute to indicate what other objects are needed to support them, particularly those objects that model physical equipment. Many also have a read-only attribute to indicate other objects that they affect.

Objects that model source trail termination functions have a read-only pointer to indicate the direction of the outflow of information and this can point to the connection termination point object modeling the source adaptation functions. The connection termination point object in turn has a read only pointer that can point back upstream to the object modeling the source trail termination point. Sink information flows are modeled in the same way and there are two sets of pointers for bidirectional flows.

Objects may have a read-only pointer attribute to an object that models a cross connection if the information flows can be cross-connected.

Objects may have read-only data attribute that indicates their operational state and a read-write attribute controls and indicates their administrative state.

Objects also have attributes to represent some of the information sent and received in the various overheads. There are also related attributes that can indicate the expected information in the overhead. A mismatch here can result in a notification being sent.

4.6 Summary

The Synchronous Optical Network (SONET) and the Synchronous Digital Hierarchy (SDH) are complex because of their hierarchical structure and ability to transport unsynchronized traffic. SONET is the version of the global SDH standards that is preferred in North America. The SONET/SDH hierarchy is based on the iterated multiplexing of 64 Kbps channels and this is reflected in its functional architecture and management model.

Three laws can be used to describe SONET/SDH. The first law governs overheads and pointers. It states that payloads are made into virtual containers (VCs) by adding a path overhead, then low-level VCs are made into tributary units (TUs) and high-level VCs made into administrative units (AUs) by adding a pointer. Path overheads are key to the management of the multiplexing layers and pointers are key to the transport of unsynchronized traffic.

The second law governs low-level payloads (1,544 Kbps and 2,048 Kbps). It states that the TUs are made from low-level payloads and are grouped into Tributary Unit Groups (TUGs) that are payloads of the high-level VCs that are made into AUs.

The third law governs the formation of interfaces. It states that groups of AUs (AUGs) are turned into SONET/SDH interfaces by adding section overheads.

These laws describe the transport of traditional plesiochronous over SONET/SDH. 1,544 Kbps (or 2.048 Kbps) payloads are made into VC-11s and TU-11s (or VC-12s and TU-12s) by adding overheads and pointers. These are made into TUG-2s, which are payloads into VC-3s or VC-4s, which are made into AUs and AUGs and so into SONET/SDH interfaces by adding the section overheads.

VC-3s are anomalous because of the historical differences between regions. A VC-3 can act as a lower-order container and be made into a TU, or it can act as a higher order container and be made into an AU.

In the functional architecture there are transport layers that consist of path layers that are served by the transmission media layers. The path layers in turn serve the circuit layers, which carry plesiochronous tributaries, and are split into lower-order and higher-order path layers to mirror the split into lower-order and higher-order containers. The transmission media layers consist of the multiplex and regenerator section layers, that map to the SONET/SDH section overheads, and the physical media layer that is specific to the transmission medium.

Trail termination functions within layers and adaptation functions between layers handle the layer overheads and pointers, respectively. The layer overheads carry operations and maintenance signals. Framing pattern overheads are handled by the adaptation functions. The components of the functional architecture are the basis of the management model that consists of interrelated objects with attributes that can be accessed, actions that can be invoked, and notifications that can be generated.

ATM is conveyed over SONET/SDH directly in VC-4s.

References

[1]   Sexton, M. and A. Reid, *Broadband Networking: ATM, SDH and Sonet*, Norwood, MA: Artech House, 1997.

[2]   *Synchronous Digital Hierarchy Bit Rates*, ITU-T Recommendation G.707.

[3]   *Synchronous Multiplexing Structure*, ITU-T Recommendation G.709.

[4]   *Network Node Interface for the Synchronous Digital Hierarchy*, ITU-T Recommendation G.708.

[5]   *Architectures of Transport Networks based on the Synchronous Digital Hierarchy*, ITU-T Recommendation G.803.

[6]   Tesink, K., *Definitions of Managed Objects for the SONET/SDH Interface Type*, RFC 2558 (http://www.ietf.org/rfc/rfc2558), March 1999.

[7]   *SDH Management Information Model for the Network Element View*, ITU-T Recommendation G.774.

[8]   *Generic Network Information Model*, ITU-T Recommendation M.3100.

5

ATM Fundamentals and Management Modeling

It's clever, but is it Art?
—Rudyard Kipling

Asynchronous Transfer Mode (ATM) provides a flexible medium for communication that is decoupled from the physical transmission. This introduction starts with the concepts of ATM cells, virtual channels (VCs) and virtual paths (VPs). This leads on to the architectural layers and functions, and the management model for ATM.

Most people would agree that ATM is clever. Some might then say that it is too clever, while others might say that it is not clever enough. The complexity of ATM specifications could support the charge that it is too clever. Some of the problems with their implementation could support the charge that it is not clever enough.

Whether or not ATM is more than just clever is left for the reader to judge.

5.1 Paths, Channels, and Cells

The basic units of ATM transmission are ATM cells [1]. These are relatively small (53 octets) in comparison to the packets used in data communications to allow them to be switched quickly by hardware and filled quickly by applications. ATM cells can be assigned to applications flexibly and they support statistical multiplexing since applications do not need to generate cells at a constant rate.

ATM cells flow over a virtual transport network specified on architectural layers that are above the physical transmission layers.

5.1.1   Physical Paths, Virtual Paths, and Virtual Channels

ATM cells are carried over physical transmission paths that are separate and distinct but may be multiplexed together so long as the identity of each transmission path is preserved. Each transmission path carries ATM cells in the same order in which they were transmitted.

ATM VPs are decoupled from the variety of physical transmission paths and physical interfaces. A VPC can pass through several ATM switches. Between each pair of switches there is a VP Link (VPL) that has a VP Identifier (VPI) that is unique and constant (see Figure 5.1). A VPC consists of a sequence of VPLs. VPIs may be changed between VPLs by VP layer switches.

VPCs are subdivided into VCs. Each VPC can carry a number of VCs that are routed together across all or part of an ATM network. In the same way that a VPC can pass through a number of VP level switches, a VCC may also pass through a number of VC level switches. Each VC Link (VCL) has a VC Identifier (VCI) that is constant and unique within its VPC, but VCIs may be changed between VCLs by VC layer switches.

The double multiplexing structure of ATM allows virtual trunk routes to be created across an ATM network at the VP layer. These VP routes can carry VC layer traffic from different end-users. The fixed size of ATM cells limits the number of ATM layers that can be defined and two layers is the minimum that allows ATM network routes to be decoupled from the physical transmission.

5.1.2   The Format of ATM Cells

The format of ATM cells is specified at physical interfaces, and is slightly different at User-Network Interfaces (UNIs) and Network-Node Interfaces (NNIs). Each ATM cell contains 53 octets [2, 3]. The first five of these form the cell header, and the remaining 48 form the payload (see Figure 5.2). The format of the payload depends on the application, but the format of the header is generic.

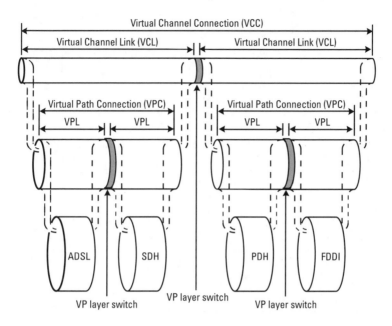

Figure 5.1 Channels, paths, connections and links.

The last octet of the header contains an eight-bit Header Error Control (HEC) field. This is a Cyclic Redundancy Checksum (CRC) that is calculated from the first four octets of the header. The CRC acts as an error correction code that allows single bit errors in the header to be corrected and multiple bit errors to be detected.

The last half of the octet that precedes the HEC contains the three-bit Payload Type (PT) field and the Cell Loss Priority (CLP) bit. The CLP bit allows cells to be tagged for early discarding (CLP = 1) or for late discarding (CLP = 0). This allows the network operator to accept cells the above the agreed data rate and to mark them for early discarding in the event of congestion.

The most significant bit of the PT field marks cells as management cells or normal cells. In management cells the other bits indicate the type and range of the management information flow. In normal cells the ATM Adaptation Layer (AAL) can use the least significant bit and the remaining bit can be used for Explicit Forward Congestion Indication (EFCI).

The VCI field consists of the sixteen bits that precede the PT field. The size of the VCI field allows over 64 thousand VCs to be multiplexed onto each VP. The total number of VPs on a physical transmission path depends

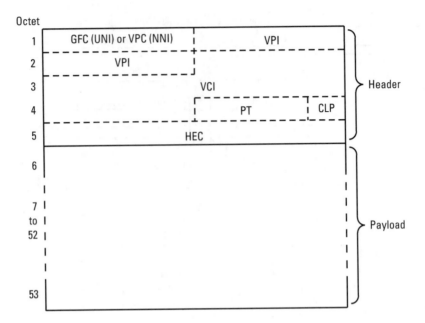

Figure 5.2   The format of cells at ATM interfaces.

on whether the path is at a UNI or an NNI. At an NNI, the VPI field consists of all 12 bits that precede the VCI field. At a UNI, the VPI field has only eight bits and is preceded by the four bits of the Generic Flow Control (GFC) field.

5.1.3   Comments on ATM Cells

The size of the VPI and VCI fields was not well thought out. Of the approximately 64 thousand values of the VCI field, often no more than sixteen are used. This was soon recognized and it has become standard practice for ATM equipment to ignore most of the VCI and VPI fields [2, 3].

It would have been more sensible to have a small VCI field for UNIs and a large field for NNIs since this would allow traffic from many users to be switched onto a single large VP at an NNI. This large VP could provide common routing for a group of connection across an ATM network and it could be overbooked since its size would allow averaging of the end-user traffic.

The GFC field that is specific to UNIs is normally set to zero to indicate uncontrolled transmission, where the network does not control the flow of cells from their UNI sources [2, 3]. When the GFC field is not zero then some form of source flow control is active. It would have been better either to omit this field because it is often unused, or to have it at both UNIs and NNIs because the flow of cells from other networks also needs to be controlled.

The length of the cell payload is a political compromise between the telecommunications and data communications worlds. A length of 32 octets was proposed in an attempt to avoid echo cancellation when ATM is used for arrowband speech. A length of 64 octets was proposed for data communications to keep overhead of the cell header low. The compromise of 48 octets is technically poor because it requires echo cancellation for voice communication, makes the cell overhead more of a burden, and is not a power of two.

5.2   ATM Layers and Functions

The physical and virtual layers of ATM can be mapped onto a functional architecture [4] with payload independent trail termination functions within layers and payload dependent adaptation functions between layers (see Figure 5.3).

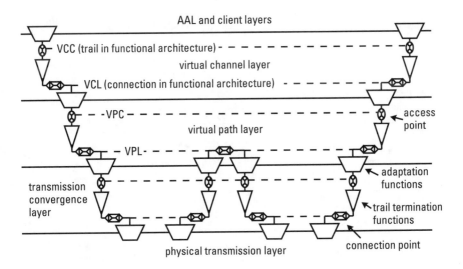

Figure 5.3 ATM layers and functions.

In the conventions used for functional architectures, the term "connection" is also used to mean a single link of an ATM virtual connection. The context should make the meaning clear.

5.2.1   The Transmission Convergence Layer

The Transmission Convergence (TC) layer inserts and deletes the idle cells that match the VP traffic to the physical transmission (see Table 5.1). It recognizes incoming idle cells by their headers since in an idle cell only the last bit of the first four header octets (the CLP bit) is one.

If there is a fixed mapping between ATM cells and the physical data stream then the TC layer can use this mapping to extract the incoming cells. If there is no fixed mapping, the TC layer identifies the cells by locking onto the pattern created by the HEC fields in the cell headers. In either case, the TC layer uses the HEC field to detect and correct header errors, and it discards cells with header errors that it cannot correct.

Table 5.1
Adaptation and Trail Termination Functions for the ATM Layers
(OAM Flows are described in more detail in Chapter 6.)

Layer	Function	Description
Virtual channel	Trail termination	End-to-end F5OAM follows
	Adaptation	Segment F5 OAM flows VC multiplexing and demultiplexing VC traffic policing
Virtual path	Trail termination	End-to-end F4 OAM flows
	Adaptation	Segment F4 OAM flows VC multiplexing and demultiplexing VC traffic policing
Transmission convergence	Trail termination	Cell header handling
	Adaptation	Scrambling Rate adaptation Cell insertion and extraction

5.2.2 The VP and VC Layers

The VP layer is the client of the TC layer. The VP adaptation functions multiplex and demultiplex the virtual connections of the VP layer. These are the VP layer trails.

The VP adaptation functions also handle the segment Operations, Administration and Maintenance (OAM) flows for the VP layer. These are the segment F4 flows. The trail termination functions of the VP layer handle the end-to-end F4 flows. Technically the VCs that carry OAM flows should not terminate in the VP layer because VCs should terminate in the VC layer. There are no VP trail termination functions in a VP layer switch because a VP layer switch only interconnects VPLs (see Figure 5.4).

The VC layer is the client of the VP layer. The VC adaptation functions multiplex and demultiplex the virtual connections, or trails, of the VC layer. VCs carry their own OAM flows, the F5 flows. The PT field in segment F5 cells has the value four. In end-to-end F5 cells it has the value five. The trail termination functions handle the end-to-end F5 flows and the adaptation functions handle the segment F5 flows.

F4 and F5 flows are described in more detail in Chapters 6 and 19.

5.2.3 Higher Order Layers

The information carried in ATM connections can take different forms, but most of it has a common characteristic. It is not organized into blocks that correspond to the 48 byte payloads of ATM cells. Between the services that use ATM and the VC layer there needs to be another layer that matches the

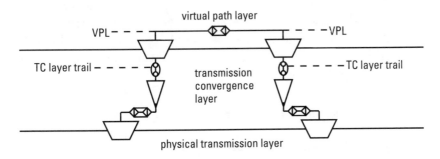

Figure 5.4 Functional architecture of a VP switch.

client services to the ATM payloads. This is called the ATM Adaptation Layer (AAL).

Five types of AAL have been defined. Each of these has a Segmentation and Reassembly (SAR) sublayer and a Convergence Sublayer (CS). The SAR sublayer performs the conversion between the client service and the ATM cell payloads. The CS is responsible for the integrity of this conversion.

Of the five types of AAL, the most commonly used are AAL1 and AAL5. AAL1 is used for fixed rate services and AAL5 is used for variable rate services. AAL3 and AAL4 were also designed for variable rate services. AAL3 was designed for connection oriented services and AAL4 was designed for connectionless packet services. Subsequently AAL3 and AAL4 were merged into AAL3/4, but this is less popular than AAL5 because it is more complex.

AAL2 could be used for variable rate video and audio services. These services normally operate at fixed data rates, but intelligent codecs can turn them into variable rate services by transmitting less information when the input signal changes slowly. AAL2 also carries the timing information for the input signal. Like AAL3/4, AAL2 is also not popular because of its complexity.

The various AALs are described in more detail in Chapter 7.

5.3   The Basic Management Model for ATM

The management model described here is the Common Management Interface Protocol (CMIP) model from ITU-T recommendation I.751 [5], which is endorsed by the ATM Forum. The Simple Network Management Protocol (SNMP) Management Information Base (MIB) for ATM [6] is not described. Although SNMP is effective for the management of small ATM switches, CMIP can scale up more easily for large switches and follows naturally from the functional architecture for ATM. The CMIP model has also be extended to include the management of on-demand connections and soft permanent connections (see Chapters 9 and 19).

5.3.1   ATM Configuration Management

5.3.1.1   Trail and Connection Termination Points

Trail Termination Point (TTP) objects represent the trail termination functions in the ATM functional architecture and Connection Termination Point (CTP) objects represent the adaptation functions. The basic model has three classes of TTP objects, corresponding to the TC, VP and VC layers.

There are also two classes of CTP objects for the adaptation functions between these three layers (see Figure 5.5).

The CTP objects are contained within the TTP objects of the layers that serve them (see Figure 5.6) because the adaptation functions cannot exist independently of the supporting trail termination functions. The CTP objects of the VC and VP layers are similar, as are the TTP objects. The TTP objects of the TC layer are different.

TC TTP objects raise an alarm if the trail termination functions that they model cannot extract cells from the incoming physical transmission. To activate or deactivate the scrambling of ATM cells on a external physical link,

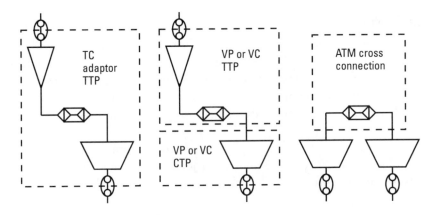

Figure 5.5 Mapping of ATM functional architecture onto managed objects.

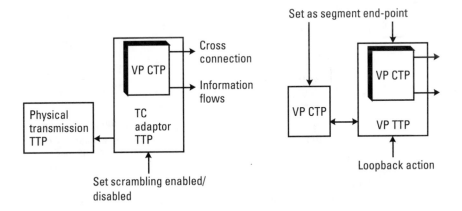

Figure 5.6 Features of trail and connection termination points.

the Operations System (OS) of an ATM switch has to set the scrambling enabled attribute in the TC TTP object for the link to true or false.

The CTP objects can describe both their normal and their OAM traffic flows. All of the TTP and CTP objects have pointers that indicate the flow of information in their layers. The CTP objects can also point to cross connection objects.

To generate loop back OAM cells at the end-point of an ATM connection the OS of an ATM switch has to send a loop back action to a VP or VC TTP object. To define that the end of a ATM link is also a segment endpoint, the OS has to set the segment end-point attribute in the VP or VC CTP object to true.

5.3.1.2 Profiles

There are often limits on the VPIs and VCIs on ATM connections and the capacity that is available for new connections is also limited. ATM access profiles describe either the profile of the VPs that are supported by a trail on the TC layer or the profile of the VCs that are supported by a trail on the VP layer. Access profiles are contained in the appropriate TTPs (see Figure 5.7).

The OS of the ATM switch perform a get on the profile object within a TTP object to determine the number of identifier bits and ATM connections that are supported and allowed.

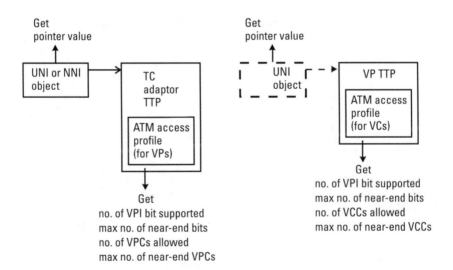

Figure 5.7 Interface objects and profile objects.

5.3.1.3 Interfaces

Interface objects point to TC TTPs to indicate whether the format of their cell headers are those of UNIs or NNIs (see Figure 5.7). There is one object class for UNIs and two for NNIs. The two NNI classes correspond to NNIs within network domains (intra NNIs) and NNIs between domains (inter NNIs) . A UNI object can alternatively point to a VP TTP if the VP TTP corresponds to a UNI that is independent of a physical interface.

Interface objects can also have a loopback location attribute. If the loopback location identifier of a loopback OAM cell (see Chapter 6) matches the value of this attribute then the OAM cell is looped back at this interface.

To discover the TTP object that the interface refers to, the OS of an ATM switch can perform a get on the pointer attribute. The OS cannot set the attribute after the interface object is created because the attribute is read only. To change the pointer the OS must delete the interface object and create a new one.

5.3.1.4 Cross Connections

Each ATM cross connection is represented by a cross connection object. These objects are contained in the fabric object that represents the cross connection matrix (see Figure 5.8).

The cross connection objects are created by an action on the fabric object. This action can also create the CTP objects at the ends of the cross connection. The fabric object also controls the flow pointers for the CTP objects.

To create a cross connection, the OS of the ATM switch sends a connect action to the ATM fabric. This action identifies the CTP objects at the

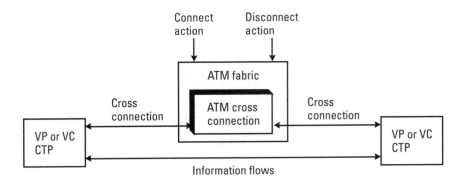

Figure 5.8 Relationships with cross connection objects.

ends of the connection or includes a description of the ATM traffic if no CTP exists. The description of the ATM traffic allows an appropriate CTP object to be created. To remove a cross connection the OS sends a disconnect action to the ATM fabric.

An ATM cross connection object points to its CTPs, which in turn point back. Each cross connection object also has an attribute that indicates whether it operates at the VP level or at the VC level.

5.3.2   ATM Performance Management

In the CMIP management model, ATM performance is monitored over 15 minute and 24 hour intervals. Current data objects hold the results of ongoing monitoring. These objects are contained in the objects that represent the places where performance is being measured. An alarm notification can be sent when the results of ongoing monitoring exceed a predefined threshold. The records of previous results are held in history data objects that are contained within the current data objects.

5.3.2.1   Interface Anomalies and Header Errors

Cells with anomalous headers are noted as records in a log. Each record notes the date, time, and VPI and VCI values of the cell, and whether the VPI and VCI values were out of range or unassigned. Each record also points to the interface that the cell came from (see Figure 5.9).

For each TC, a count is kept in a current data object of the number of cells with header errors and of the number of cells that have been discarded because these errors were uncorrectable. For each interface, a count is kept of the total number of cells that have been discarded due to uncorrectable header errors or anomalous VPI or VCI values.

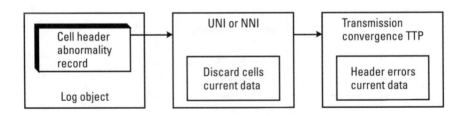

Figure 5.9   Interface anomalies and header errors.

5.3.2.2 Traffic Monitoring and Traffic Policing

Both end users and operators of other networks can exceed their agreed traffic levels. The policing of user traffic is called Usage Parameter Control (UPC) and the policing of network traffic is called Network Parameter Control (NPC) [7]. The adaptation functions between ATM layers can count the number of cells that they pass and discard. ATM switches can also count the number of incoming and outgoing cells at external interfaces and at the adaptation functions between the layers.

These cell counts are kept in traffic monitoring current data objects that are contained in either the interface objects or in the CTPs. The CTPs also contain current data objects that count the number of cells that are changed from a low loss priority (CLP = 0) to a high loss priority (CLP = 1).

5.3.2.3 Performance Monitoring Flows

Although OAM cells are described in more detail in a Chapter 6, the basic modeling for performance monitoring (PM) OAM cells is described here because this is part of modeling of performance management.

The adaptation and trail termination functions can generate and process both segment and end-to-end OAM flows. The performance monitoring objects that model this functionality are contained in the appropriate TTP and CTP objects. Performance monitoring objects have attributes that indicate the average number of cells in a performance monitoring block and whether generation and processing is active.

Performance monitoring objects also have an action to request either active generation and termination of cells or just passive monitoring. The action also specifies the desired behaviour at the far end of the performance monitoring flow. Performance monitoring objects can also indicate the direction and nature (segment or end-to-end) of the flow and whether the far end is active.

If the adaptation or trail termination functions are processing PM flows then they keep counts of the number of ordinary cells and the numbers of lost and misinserted cells. They keep these counts for the local processing and can also hold the counts of the far end processing if this is reported back. These counts are modeled as current data objects that are contained in the performance monitoring objects.

5.4 Summary

ATM creates a logical medium for communications that is decoupled from the physical transport that supports it. This logical medium is asynchronous as the time taken for an ATM cell to traverse it is not fixed. The transmission paths that carry ATM cells vary with the media and techniques used for physical transmission, but the structure of the ATM cells is independent of the physical transmission.

Each physical transmission path that carries ATM cells supports a number of Virtual Paths (VPs), and each VP carries a number of Virtual Channels (VCs). The VP and VC of each ATM cell within a transmission path are identified by the value of the VP and VC Identifiers (VPI and VCI) in its five-octet header. The cell's payload is the remaining 48 octets in the cell.

The VPI field in the header of ATM cells that belong to a Network-Node Interface (NNI) is four bits larger than for cells belonging to a User-Network Interface (UNI). In a UNI these bits are reserved for Generic Flow Control (GFC) but are normally coded as zeroes.

The conversion between the physical transport and the ATM cells is handled by the Transmission Convergence (TC) layer of the ATM functional architecture. The TC layer serves the VP layer that in turn serves the VC layer. The final application layers are clients of the ATM Adaptation Layer (AAL) that the VC layer serves.

The TC functions handle cell delineation that depends on the physical transport, and cell Header Error Check (HEC) generation and validation that is independent of the physical transport. The VP and VC trail termination functions handle the VP and VC layer operational overheads that are independent of the layer payloads, while the adaptation between the VP and VC layers handles the ATM Payload Type (PT) field of the cell headers.

Configuration of ATM equipment can be performed using a management model that is based on the functional architecture. The functions performed at the various levels are modeled as managed objects and objects are also used to differentiate between UNIs and inter- and intra- NNIs. Profile objects indicate the number and nature of the VPs that are allowed on a physical transmission path and the VCs that are allowed on a VP.

ATM cross connection can be performed at the VP layer or at the VC layer. Objects are used to model the individual cross connections, and these are controlled by a fabric object that represents the complete cross connection matrix.

The TC functions keep performance statistics about cell errors and cells with invalid VP and VC identifiers. The interfaces keep count of cells passed. The adaptation functions between ATM layers also keep count of cells passed and of cells that are discarded due to traffic policing. Counts of lost and misinserted cells are detected at the ends and intermediate points of ATM paths by the performance monitoring OAM flows.

References

[1]   de Pryker, M., *Asynchronous Transfer Mode Solution for Broadband ISDN*, New York: Ellis Horwood, 1993.

[2]   McDysan, D. E., and D. L. Spohn, *ATM Theory and Application*, New York: McGraw-Hill, 1995.

[3]   Sexton, M., and A. Reid, *Broadband Networking*, Norwood, MA: Artech House, 1997.

[4]   *Functional Architecture for Transport Networks based on ATM*, ITU-T Recommendation I.326.

[5]   *Asynchronous Transfer Mode (ATM) Management of the Network Element View*, ITU-T Recommendation I.751, March 1996.

[6]   Tesink, T., *Definitions of Managed Objects for ATM Management*, RFC 2515 (http://www.ietf.org/rfc/rfc2515). February 1999.

[7]   *Traffic Control and Congestion Control in B-ISDN*, ITU-T Recommendation I.371, August 1996.

6

ATM Operations, Administration, and Maintenance (OAM) Flows

Eighty-five percent confusion and fifteen percent commission.
—Fred Allen

There are certain myths that one must respect concerning the OAM flows for ATM. The first is that there are three layers of OAM flows for the underlying physical transmission. This is why the lowest layer of the ATM flows is called the F4 layer. The F1, F2, and F3 flows are defined for SDH transmission, and since ATM is supposed to be decoupled from the physical transmission we must pretend that in this respect all forms of physical transmission are like SDH.

We must also pretend that the VC connections that carry the F4 flows are not really VC connections at all. This is because the F4 flows belong to the VP layer of ATM, and if they were carried in VC connections they could not terminate in the VP layer. This is the second myth that one must respect.

The third myth is the most difficult. We must also pretend that suppliers of ATM equipment are eager to implement OAM flows. If we do not respect this myth then, without the benefit of chemically induced enlightenment, it is difficult to understand why so much work effort has already been expended.

At least we can be confident that the numerous shortcomings in the defined flows and the level of confusion ensure that suppliers who do implement them will have earned their commission.

6.1 OAM Flow Layers and Ranges

ATM transmission is supported by the physical transmission, and each of these is divided into layers, two for ATM and three for physical transmission (see Figure 6.1). The layers of the physical transmission are derived from the functional architecture for SDH. The highest physical transmission layer, the transmission path layer, supports the ATM VP layer. The F1 to F3 OAM flows are associated with the three physical transmission layers. The F4 flows are associated with the VP layer and the F5 flows are associated with the VC layer [1].

One or more adjacent ATM links can be joined to form a segment. Typically a segment is owned by a network operator and can only cross the

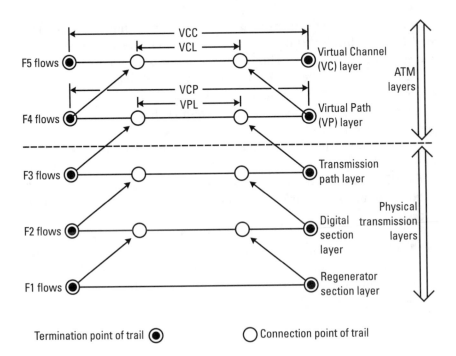

Figure 6.1 Transmission layers and OAM flows.

domains belonging to different operators by agreement. OAM flows may belong either to the entire ATM connection or to a particular segment (see Figure 6.2). Those belonging to the entire connection are called end-to-end OAM flows and those belonging to a particular segment are called segment flows. Segment flows must not cross segment boundaries, since they could then interfere with the segment flows of another operator.

The layers and ranges of the cells used for OAM flows are indicated in the fields in their headers (see Figure 6.3). F4 flows are identified by the value of the VCI field. Segment F4 flows have a VCI value of three, and end-to-end F4 flows have a value of four. Likewise, the F5 flows are identified by the value of the PT field. Segment F5 flows have a PT value of four and end-to-end flows have a value of five.

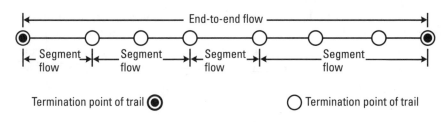

Figure 6.2 Ranges of ATM OAM flows.

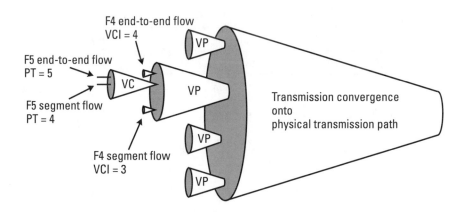

Figure 6.3 F5 and F4 flows.

6.2 OAM Cell Types and Functions

The first octet of the payload of an OAM cell defines its type and its function (see Figure 6.4). The first four bits of the octet indicate one of four OAM cell types. The remaining four bits indicate the function within that cell type.

The two main types of OAM cells are fault management (FM) cells and performance management (PM) cells (see Table 6.1). PM cells can carry an error detection code for blocks of normal cells so that their error rate can be monitored. They also allow the result of the monitoring to be reported back.

The third type of OAM cell is used for activation and deactivation. This is a subsidiary type that is used to support FM and PM functions by activating or deactivating performance monitoring by PM cells and continuity checking by FM cells. The fourth and final type of OAM cells are System Management (SM) cells. The use of SM cells is specific to particular implementations of ATM systems and is not standardized.

All but the last two octets of an OAM cell payload can be used for function specific fields. Unused intermediate octets are coded as 6A in hexadecimal notation. The last two octets contain six reserved bits that are fixed at zero and a ten bit Cyclic Redundancy Checksum (CRC) that is calculated for all of the previous bits of the payload.

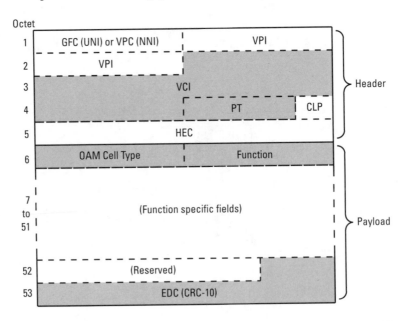

Figure 6.4 The format of ATM OAM cells. Shading indicates generic fields used by all OAM cells.

Table 6.1
Octet 6: OAM Type and Function

Cell Type	Function	Description
0001—Fault management	0000—AIS	Ongoing alarm indication signal
	0001—RDI	Returned defect indication
	0100—Continuity check	Maintains minimum cell flow
	1000—Loopback	Non-intrusive loopback cell
0010—Performance management	0000—Forward monitoring	Ongoing bit interleaved
	0001—Backward reporting	Returned performance report
1000—Activation/ deactivation	0000—Forward and backward performance monitoring	Activates and deactivates monitoring and reporting
	0001—Continuity check	Activates and deactivates continuity checking
1111—System management	(not specified)	(system specific use)

6.2.1   Fault Management

FM cells can perform one of four functions (see Figure 6.5). When a fault is detected, Alarm Indication Signal (AIS) cells are sent onwards on the affected connections to inform the downstream nodes that a problem exists. Upstream nodes are informed of the fault by the return of Remote Defect Indication (RDI) cells.

The remaining two functions that FM cells can perform are loopback and continuity checking. Continuity Check (CC) cells are inserted when the normal traffic on a connection drops below a certain rate. CC cells confirm that the connection still exists even when it is carrying little or no traffic. Loopback cells confirm that there is some form of bidirectional connection to the location that they are looped back from.

6.2.1.1   Alarm Indication Signal (AIS) and Remote Defect Indication (RDI)

The AIS and RDI functions are closely related since they both perform defect indication. The different names are for the two different directions, ongoing (AIS) or returned (RDI). There are two function specific fields for defect indication (see Figure 6.6), the defect type field and the defect location field. Both of these fields are optional, and in practice they have been ignored

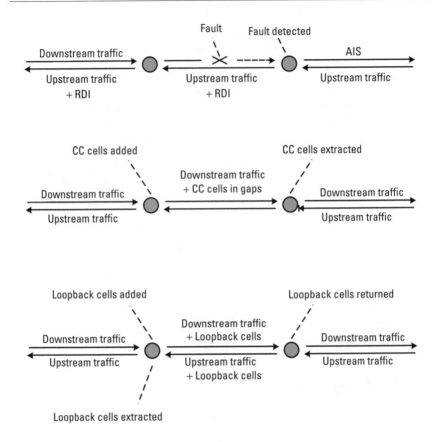

Figure 6.5 Fault management flows for ATM.

although this may change with implementations of the 1999 version of I.610 (see Chapter 19).

6.2.1.2 Continuity Check (CC) Cells

In CC cells, all of the 45 octets that are available for function specific fields are unused. This means that the payload of a CC has a constant value. An advantage of this is that it makes implementations easier because it is not necessary even to calculate the CRC for the payload. The disadvantage is that the potential of using CC to uniquely label connections is lost.

6.2.1.2 Loopback Cells

Loopback testing using OAM cells is different from conventional loopback testing because it does not interfere with the normal traffic on the

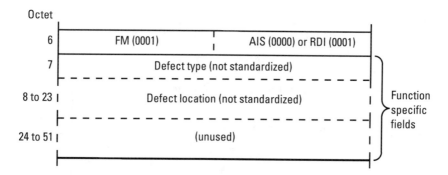

Figure 6.6 Function specific fields for AIS and RDI.

connection. This is because only the loopback cells are returned, not the normal traffic cells. This is not the advantage it first appears because it means that a loopback test is likely to give a positive result on a degraded connection until the degradation becomes severe.

There are four function specific fields for loopback cells (see Figure 6.7). The first of these is the mandatory loopback indication field. All of the bits of the loopback indication field except the least significant bit are fixed at zero. The least significant bit indicates whether or not the cell has been looped back. When a cell is first generated the least significant bit is set to zero. When the cell is looped back, this bit is set to one.

The next field is the mandatory correlation tag that is used to identify returned cells. The subsequent field is the mandatory loopback location ID that identifies the node where the cell is to be looped back. The last field is the optional source ID that identifies the origin of the loopback cell. Unfortunately the values of these three fields are not standardized and so the fields are ignored. In practice the payload of loopback cells can have two values, one value for generated cells and one for returned cells.

6.2.2 Performance Management

The two functions performed by PM cells are forward monitoring and backward reporting (see Figure 6.8). Forward monitoring cells carry a field that holds the parity checksum calculated for the block of traffic cells. Backward reporting cells inform the node sending the forward monitoring cells of the number of errors in each monitored block. When a backward reporting cell is sent, it is paired to a particular forward monitoring cell.

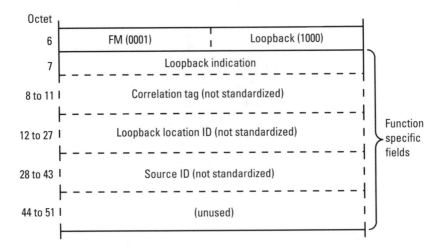

Figure 6.7 Function specific fields for loopback.

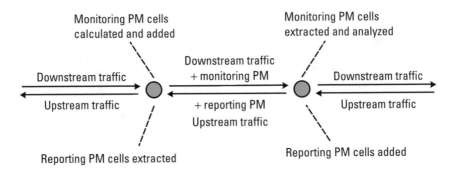

Figure 6.8 Performance management flows for ATM.

Backward reporting via PM cells is not strictly essential because the result of performance monitoring could be reported through the Operations Systems (OSs) of an ATM network.

6.2.2.1 Common Fields

There are four function specific fields for PM cells that are common to both forward monitoring and backward reporting (see Figure 6.9). All of these fields are mandatory in PM cells, but one of them, the time stamp field, is allowed to take a default value.

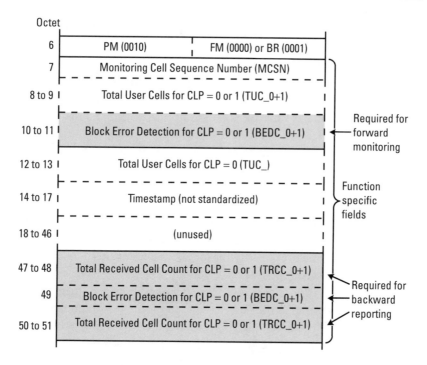

Figure 6.9 Function specific fields for performance monitoring. Shading indicates fields that are specific to forward or backward PM cells.

The second octet of the payload of a PM cell holds the monitoring cell sequence number (MCSN). This field numbers PM cells in sequential order so that it is possible to determine if any intermediate cells have been lost. Independent counters are needed at either end so that cells lost in either direction can be detected. Co-ordination of the sequence numbering is achieved through activation and deactivation cells.

The next payload field holds a count of the total user cells transmitted. The difference in the value of this field between two transmitted forward monitoring cells gives the size of the block that has been monitored. The value of this field in a backward reporting cell is copied from the value in the forward monitoring PM cell to which it is paired so that it can be matched to the transmitted forward monitoring cell. This approach lets performance monitoring skip on when PM cells are lost.

There is a second field that also carries a count of user cells. This is a count of high priority cells that have a CLP bit set to zero. It follows a field that is only used for forward monitoring cells.

The last of the four fields that are common to both forms of PM cell is the time stamp field. This field can be used to indicate the time when the PM cell was inserted. One of the advantages of time-stamping is that it allows the variation in cell arrival times, the cell delay variation (CDV), to be measured. Unfortunately the way that the field should be used is not standardized. The result is that the default value of all ones is normally used and the possibility of measuring CDV is lost. This field must be present to prevent the normal coding for unused octets from being misinterpreted as a timestamp.

6.2.2.2   Forward Monitoring Specific Field

The single field that is specific to forward monitoring cells is the mandatory block error detection field (see Figure 6.9). This field carries result of the Bit Interleaved Parity (BIP) calculation for a block of user cells. Each bit of the sixteen bits in this field is a parity bit for a sequence of user bits spaced sixteen bits apart, i.e. bits 1, 17, 33,...; 2, 18, 34,...; etc.

If there is no more than one error in each sequence, then the total number of parity errors in the sixteen sequences is the total number of errors in the block. If a block is completely corrupted then on average errors will be detected in half of the sequences.

6.2.2.3   Backward Reporting Specific Fields

There are three mandatory fields that are specific to backward reporting cells. One of these reports the result of the remote error monitoring (see Figure 6.9). Normally the value of this field is the number of parity errors (0–16) that the remote end has detected, but if the monitoring is invalid then all the bits of this field are set to one. The monitoring is invalid if the forward monitoring cells received at the remote end were not numbered sequentially or if the number of received user cells does not match that indicated by the forward monitoring cells.

The other fields in the backward reporting cells allow inconsistencies in the number of received cells to be reported. This is achieved by the remote end counting the number of received cells, both the total and the number with a CLP bit of zero. When these are reported back, the node that generated the forward monitoring cells can determine the number of cells of each category that have been lost.

6.2.3   Activation/Deactivation

Activation/Deactivation (A/D) OAM cells support PM cells and CC cells by requesting the remote node to remove (sink) a flow of PM or CC cells

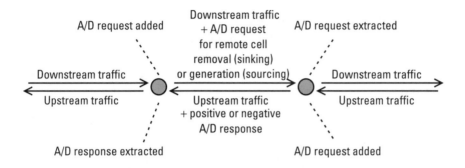

Figure 6.10 Activation/Deactivation OAM flows.

sent to it and to process them, or to generate (source) a flow in the return direction.

6.2.3.1 Common Fields

There are three function specific fields that are common to all A/D cells (see Figure 6.11). For continuity checking, no other function specific fields are needed.

The second octet of the payload contains two function specific fields. The message ID field consists of the first six bits of the octet and remaining two bits form the directions of action field. Six values are defined for the message ID field, three for activation and three for deactivation. In each case there is a command and two responses, one positive and one negative.

The directions of actions field indicates the direction or directions of the controlled flow. A command that specifies the away (or A-B) direction is

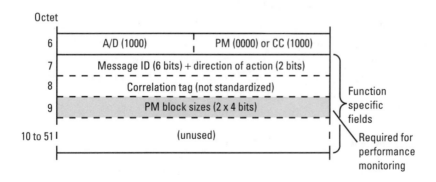

Figure 6.11 Function specific fields for activation and deactivation.

a command to activate or deactivate the sinking of cells because the direction of the flow is the same as that of the command. A command that specifies the return (or B-A) direction refers to the remote sourcing of cells. Both directions may be specified if the command is to activate or deactivate both sinking and sourcing, and neither direction need be specified if directionality is not relevant.

The third common field is the correlation tag. This is intended to allow responses to be associated with commands. Unfortunately the guidance on the use of the correlation tag is not especially helpful.

6.2.3.2   Performance Monitoring Specific Fields

The two performance monitoring specific fields in the fourth payload octet are the four-bit block size fields, one for each direction. These fields allow negotiation of the nominal block sizes to be used for performance monitoring. The block sizes are nominal because forward monitoring PM cells should only be inserted when a spare cell location occurs.

These two fields are only used when the message field has the value activate or activation confirmed. In all other cases a default value of all zeroes is used.

6.3   The Operations System (OS) Interface

In theory, the OS interface for management and control of ATM OAM flows should conform to ITU-T recommendation I.751 [2] that defines the appropriate Q3 interface for the Telecommunications Management Network (TMN). For each of the OAM capabilities that are supported, the relevant actions, notifications and attributes are described.

Actions are requests or commands. They often include parameters and provoke a reply. Attributes contain information that can be read and some can be written to. Notifications are generated spontaneously. For OAM flows ITU-T Recommendation I.751 only specifies generic notifications that are not specifically tailored to ATM (see Figure 6.12).

6.3.1   The Notification for Defect Reporting

Defects are reported by the objects that represent the intermediate points and end-points of ATM connections. The generic notification that is used is able to indicate the difference between an AIS defect and an RDI defect, but it

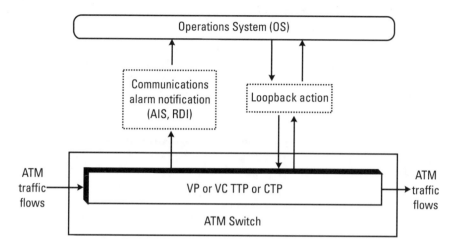

Figure 6.12 OS transactions for defect reporting and loopback.

does not provide ATM specific information from the defect location and defect type fields of the ATM cells.

6.3.2 Loopback

Loopback differs from continuity checking and performance monitoring because it is a complex OAM function that is modeled using an optional action that is added directly to the objects that represent the intermediate points or end-points of ATM connections.

The action that controls loopback is a request with a complex parameter that gives information about the loopback location and the whether the cells to be generated are for end-to-end or for segment loopback. The loopback location can be specified in terms of its loopback location field and whether or not it is a connection end-point.

The reply to the action indicates whether or not the loopback was successful.

6.3.3 Continuity Checking

Continuity checking is modeled using objects that are contained within the objects used to represent intermediate points or end-points of ATM connections. The actions, notifications and attributes for continuity checking refer to these contained objects (see Figure 6.13).

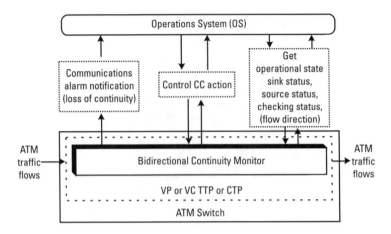

Figure 6.13 OS transactions for continuity checking.

The principal notification used for continuity checking is a generic notification that is able to report a loss of continuity.

6.3.3.1   The Continuity Check Action

The action that controls continuity checking is a request from the OS to the ATM switch that has a parameter that indicates whether or not the node should actively source or sink CC cells. This action allows the OS to directly activate and deactivate continuity checking.

The reply to the action indicates whether or not the node has responded by sourcing or sinking CC cells. The reply also indicates problems, in particular whether sinking or sourcing had already been activated and if there is no sink or source flow.

6.3.3.2   Continuity Check Attributes

There are four attributes associated with continuity checking. All of these are read-only, and all but one are mandatory. The nonmandatory attribute indicates whether the OAM flow extends from the node into the heart of the ATM equipment or outwards towards the periphery of the ATM equipment. This attribute is present except for end-points of ATM connections since there only one direction is possible.

Two of the other attributes indicate whether or not the sink and source mechanisms for OAM flows are active. The fourth attribute is the operational state for the continuity checking, which can be enabled or disabled and for which there is an associated notification.

6.3.4 Performance Monitoring

Like continuity checking, performance monitoring is also controlled using objects that are contained within the objects used to represent intermediate points or end-points of connections (see Figure 6.14). The notifications generated for performance monitoring are all related to changes in the attributes of the contained objects.

6.3.4.1 The Performance Monitoring Action

The action that controls performance monitoring is similar to that which controls continuity checking. It is a request with a parameter that determines how the node should perform forward monitoring and backward reporting. The action is more complex than that for continuity checking because, in addition to generating forward monitoring PM cells, nodes can also generate backward reporting PM cells and process either type of PM cell.

    The reply to the action is also similar to that for continuity checking in that it confirms how the node has responded. The reply can also indicate a greater range of problems than the reply to the continuity check action to reflect the greater functionality associated with PM cells.

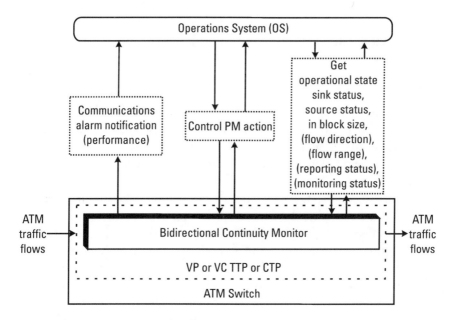

Figure 6.14 OS transactions for performance monitoring.

6.3.4.2   Performance Monitoring Attributes

All of the performance monitoring attributes are read-only. Five of these attributes are always present, and the remaining four are conditional. Three of the five correspond to the mandatory attributes for continuity checking, namely the indications of the operational state of the object and of whether or not sinking and sourcing are active. The additional two attributes indicate the average block sizes for performance monitoring on the incoming and outgoing flows.

One of the remaining four conditional attributes, the flow direction, is the same as for continuity checking and is not present at the end-points of connections where the flow can have only one direction. One of the other conditional attributes indicates whether or not the far end is monitoring the performance and another indicates whether or not the far end is reporting the result back. The final conditional attribute indicates whether the performance monitoring flow is segment or end-to-end and this is really a mandatory attribute since it must be present if the monitoring is non-intrusive. Intrusive performance monitoring is exceptional because it interferes with the monitored traffic.

6.4   Problems and Deficiencies

There has been considerable discussion about the deficiencies in the OAM flows defined in ITU-T Recommendation I.610. Some of this has been uncharitable and has suggested that certain equipment suppliers have been less than positive about correcting the deficiencies. In the interests of mental health it may be better to adopt a less paranoid outlook and to attribute the problems to human incompetence rather than to intelligent scheming.

The defect indications give similar information to continuity checking since both indicate when a connection has been lost. Continuity checking is the more effective approach because it gives a positive indication that the connection exists. In contrast, the absence of a defect indication does not mean that the connection is intact because the defect indication could have been lost. The main advantage that defect indications have over continuity checking is that defect indications have function specific fields that can provide additional information. Unfortunately these fields were not well specified in the 1996 version of I.610.

The approach to continuity checking is not ideal because continuity check cells are not labeled. This means that continuity checking will give a positive result when there is a misconnection because there is no way to distinguish between cells that should have been received and the cells that have been received due to the misconnection.

Loopbacks are less effective than might be thought because they provide no real information about the quality of connections. Performance monitoring provides a much better indication and can be used instead of loopback testing when backward reporting is included.

Performance monitoring ought to be a stand-alone capability. This is not how it is seen at present because the advice in ITU-T recommendation I.610 is to use it in conjunction with continuity checking so that it is clear that the connection is available while it is being monitored.

The nature of the performance monitoring carried out by OAM cells has also been questioned. The BIP approach gives an immediate indication of the number of errors in each block when this is low and allows a good estimate of the number of errors to be calculated when the number of errors is high. Unfortunately, this is not especially helpful because of the nature of the traffic.

If the connection is used to carry data packets that are retransmitted when corruption is detected, then it does not matter how many errors there are in a package because it will be retransmitted unless it is error-free. If the connection is used for broadcast video with bit interleaved forward error correction then the error rate does not matter until it becomes high because the traffic is immune to low error rates. In either case it would be more helpful to know the number of cells that had been corrupted or when a second of transmission had been free of errors rather than the precise number of errors in a block.

The interface to the operations system specified in ITU-T Recommendation I.610 is rudimentary. There is no way to configure the locations used in the OAM cells or to report the location or type of defects or whether a defect is indicated by an en-to-end or by a segment flow. There is a limited ability to support the results of performance monitoring that does not accurately reflect the capabilities of the PM cells. There is no way to know if continuity checking is segment or end-to-end. There is also no ability to trigger activation/deactivation cells.

Some of these deficiencies have been addressed in the 1999 version of I.610 (see Chapter 19) but in many cases no agreement on the corrections for acknowledged deficiencies seems possible likely, partly because of the reluctance of suppliers to change their implementations.

6.5 Summary

Operations, Administration and Maintenance (OAM) flows are overheads on the normal ATM traffic that provide information about ATM faults and performance and allow certain tests to be carried out. The two layers of ATM transmission, the VP and VC layers, each have OAM flows and these flows can range overt the entire ATM connection or be confined to particular segments.

The two main types of OAM cells are fault management (FM) cells and performance management (PM) cells. There is also a subsidiary type of OAM cell used for remote activation and deactivation of certain PM and FM flows. The final type of OAM cells, system management (SM) cells, is not standardized.

FM cells can carry defect indications, both ongoing (AIS) and returned (RDI). They can also perform loopback and continuity testing. PM cells can perform forward monitoring and backward reporting of performance. Activation and deactivation cells support performance monitoring and continuity checking.

The interface to the Operations System (OS) supports OAM flows through actions, notifications, and readable attributes. Defect reporting is only supported by a notification, and loopback testing is only supported by an action. All three types of transactions are used to support performance monitoring and continuity checking.

A number of deficiencies have been found in the defined OAM flows for ATM, but despite this there has been considerable resistance to efforts at correction. The deficiencies are due to incomplete standardization of the fields with OAM cells, poor understanding of requirements, and poor support of the flows at the OS interface.

References

[1]   *B-ISDN Operations and Maintenance: Principles and Functions,*
       ITU-T Recommendation I.610, 1995.

[2]   *Asynchronous Transfer Mode (ATM) Management of the Network Element View,*
       ITU-T Recommendation I.751, March 1996.

7

ATM Adaptation for Client Services

The past is a foreign country: they do things differently there.
—L. P. Hartley

ATM decouples the services that it carries from the physical transmission systems that carry the ATM cells. The ATM Adaptation Layer (AAL) is the name given to the mediation layer between the client services and their ATM connection.

It has been suggested that in the future, client services will be designed to use an ATM connection directly without the need for an adaptation layer. Whether or not this is so, the services that have been defined in the past are not able to use ATM directly. As for the future, the losers in the defined AALs indicate that the future too is a foreign country.

7.1 Introduction

The ATM Adaptation Layer (AAL) [1] matches the services carried by ATM to the payloads carried by ATM cells. Two AALs have now emerged as the clear favorites. AAL1 is the favorite for services like conventional audio and video that need a real-time synchronization signal. AAL5 [2] is the favorite for services like packet data communications that do not need the transport layer to carry a synchronization signal

AAL1 is used for video-on demand service and for the emulation of narrowband 64 Kbps circuits. AAL5 is used to carry the Internet Protocol (IP) [3].

AAL2 was also defined for services that need a synchronization signal. The transmission rate for AAL1 is constant while the transmission rate for AAL2 is variable. AAL2 is suited to variable rate codecs that transmit less data when their inputs change less. AAL1 is preferred to AAL2 because it is simpler and because it is just as effective when service layer compresses the audio or video signal.

AAL3 was defined for connection-oriented packet data communications and AAL4 was defined for connectionless packet data communications. The two were subsequently merged into AAL3/4. AAL5 was then developed because AAL3/4 seemed complex and difficult to implement, and AAL5 became the preferred approach. The sublayers of ATM adaptation handle the differences between connection-oriented and connectionless services.

7.1.1   The Structure of the ATM Adaptation Layers

The five AALs are each divided into two sublayers. The Segmentation and Reassembly (SAR) sublayer forms the lower part of each adaptation layer (see Figure 7.1). The SAR sublayer breaks the service traffic down into ATM payloads and rebuilds the payloads back up into service traffic. The SAR sublayer also handles sequence numbering and error monitoring if the AAL requires these.

The SAR supports the Convergence Sublayer (CS). The lower part of the CS, the Common Part of the Convergence Sublayer (CPCS), is defined for all AALs. The CPCS handles the errors and problems that are detected by the SAR sublayer. It also handles the synchronization signals for AAL1 and AAL2.

A higher part of the CS, the Service Specific Convergence Sublayer (SSCS) that is supported by AAL5 has been defined for call control signaling over ATM interfaces (see Figure 7.1). The lower part of this SSCS, the Service Specific Connection-Oriented Protocol (SSCOP) [4], is common to both the User-Network Interface (UNI) and the Network-Node Interface (NNI). The higher part, the Service Specific Co-ordination Function (SSCF), takes a different form for UNIs [5] and for NNIs [6]. The combination of the service independent AAL5 and the SSCS is called the Signaling AAL (SAAL) [7].

Other SSCFs have also been defined. These include SSCFs for the connection-oriented Frame Relay service and for the connectionless

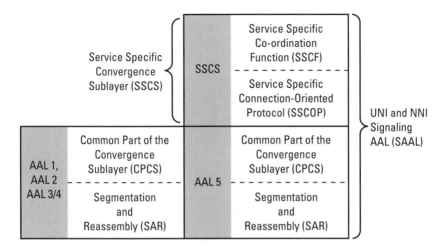

Figure 7.1 The structure of the ATM adaptation layers.

Switched Multimegabit Data Service (SMDS). There is no SSCF or SSCS defined for IP because IP has its own way of providing this functionality.

In practice the most common applications keep the functionality of the AALs to a minimum. The introduction of call control signaling, with its SSCS, has been slower than anticipated. In contrast, ATM transport of IP, which does not need an SSCS, is a well established. The differentiation between the process of segmentation / reassembly and the validation of this process makes sense, but the unnecessary complexities introduced for AAL3/4 (see Figure 7.2) and AAL2 have been their undoing.

7.1.2 Management Modeling of ATM Adaptation

The management model for ATM adaptation is specified in ITU-T Recommendation Q.824.6 [8]. This was agreed between the ITU-T and the ATM Forum and was endorsed by the European Telecommunications Standards Institute (ETSI).

Profile objects represent the characteristics of the various adaptation layers (see Figure 7.3). Each profile has an attribute that indicates the AAL to which it refers and a set of attributes that are particular to that AAL. A second type of profile represents the characteristics of circuit emulation over AAL1. Current data objects represent the active performance monitoring of ATM adaptation. Each current data object has a set of attributes that is particular to a specific AAL.

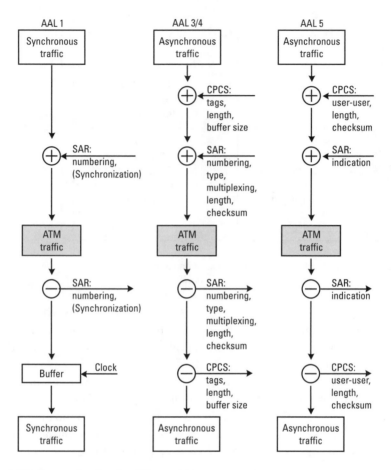

Figure 7.2 Processing flow for different AALs.

For example, to change the AAL parameters at single VCC endpoint the Operations System (OS) writes to the interworking VC TTP object to change its AAL profile pointer to refer to a new profile. To define a new profile, the OS has to create a new AAL1 profile object that the VC TTP objects can refer to.

7.2 Synchronous Traffic (AAL1 and AAL2)

AAL1 is used by synchronous services that require a synchronization signal and a constant flow of data to be carried across an ATM network. Although

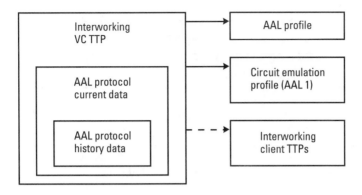

Figure 7.3   The management model for ATM adaptation and circuit emulation.

it is simpler to carry synchronous services over a synchronous network, ATM networks also need to be able to carry synchronous services and the flexibility of an ATM network avoids the need for a separate network for each synchronous service. Synchronous services often require a buffer to compensate for the Cell Delay Variation (CDV) experienced by the ATM cells.

7.2.1   AAL1 Segmentation and Reassembly

In AAL1, the first byte of the payload of each ATM cell is the SAR header (see Figure 7.4). The sequence count field in the SAR header numbers the ATM cells cyclically from zero to seven. A sequence length of eight cannot measure a loss of eight or more consecutive cells, but the length is sufficient because such a loss is detectable as an interruption in the synchronous traffic.

The SAR sublayer also generates an error correction field from the SN field (see Figure 7.4). When the SAR sublayer receives cells it uses this field to correct single bit errors in the SAR header. The additional parity bit provides a means of detecting errors that cannot be corrected. The SAR sublayer should generate an alarm if detects an uncorrectable error, or an error in the cyclic numbering, or an extended absence of cells.

The four bits of the SN field and the three bits of the error correction field form a seven bit Hamming code [9] that can correct a single bit error. A single bit error can be corrected because adding three bits allows one of eight situations to be identified. The eight situations are the identification of one of the seven bits that has been corrupted or the confirmation that none of the seven bits have been corrupted. If two of the seven bits have been corrupted then the corrupted code resembles a different code that has only one bit corrupted, but this is contradicted by the parity bit.

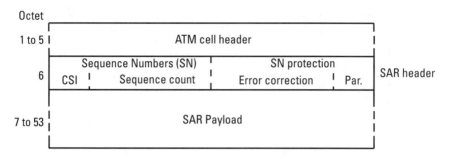

Figure 7.4 Segmentation and Reassembly (SAR) for AAL1.

The AAL1 CPCS can also use the SAR header in the regeneration of clock and frame synchronization.

7.2.2 AAL1 Synchronization Modes

The CPCS of AAL1 handles the synchronization signal for constant rate services. The Synchronous Residual Time Stamp (SRTS) method conveys clock synchronization. The Structured Data Transfer (SDT) method conveys frame synchronization. The two methods can be used independently or together.

7.2.2.1 Clock Synchronization: The SRTS Method

The SRTS method allows a clock to be generated at the receiving end of a connection that is synchronized to a clock at the transmitting end, regardless of CDV. The transmitting end counts reference clock cycles, but only holds the least 4 significant bits. These four bits are the residual time stamp and they provide enough information to fine tune a local reference clock at the receiving end.

The transmitting end reads the counter each time eight ATM cells have been filled with payload bits and sends the residual time stamp to the receiving end (see Figure 7.5) in the CSI bit in the SAR headers of the four ATM cells with odd AAL1 sequence numbers. The CSI bit in the four cells with even sequence numbers is zero.

The receiving end can generate a local residual time stamp from its local clock and can adjust the speed of its local clock so that the local residual time stamp tracks the residual time stamp that it receives from the transmitting end.

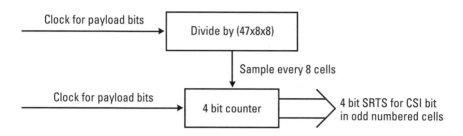

Figure 7.5 Synchronous Residual Time Stamp (SRTS) generation.

7.2.2.2 Frame Synchronization: The SDT Method

The SDT method conveys frame synchronization across an ATM connection. Seven of the eight bits of the octet that follows the SAR header of the even numbered AAL1 cells can carry an SDT pointer that indicates a boundary between two payload octets that corresponds to a frame boundary.

If a frame boundary does not start within the 93 octets that follow the pointer then the value of the pointer is all ones. Otherwise the value of the pointer is in the range zero to 92 (see Figure 7.6). A phase-locked loop can generate a bit clock for the service that is locked to the frame boundaries.

It would have been possible for the SDT pointer to be carried in the CSI bit of the SAR headers but the frame boundaries could not then be as close and it would prevent the SDT and SRTS methods being used together.

7.2.3 The AAL1 Convergence Sublayer (CS)

The AAL1 CS is often used for circuit emulation, i.e. to emulate a fixed rate physical circuit. Circuit emulation normally needs a CDV buffer that may overflow or underflow.

If audio or video services are compressed, they become more sensitive to corruption from errors in the transmission. The AAL1 CS can add Forward Error Correction (FEC) to compensate for this. FEC is more effective if it operates over long blocks of data, but this increases the overall transmission delay. If the service is video-on-demand then an increased transmission delay is less significant because the level of interaction in the service is low. If the service is voice telephony then it is more sensitive to delay because it is more interactive.

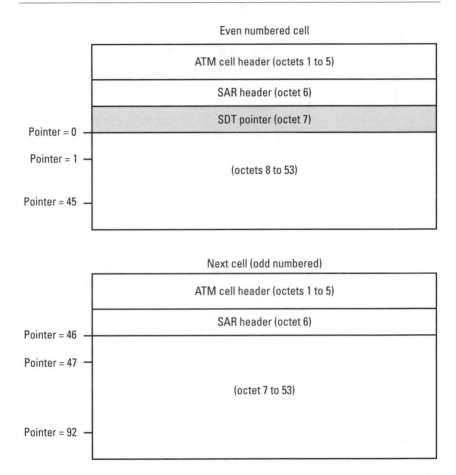

Figure 7.6 Structured Data Transfer (SDT): frame boundary indication.

7.2.4 AAL1 Management and Modeling

The configuration of AAL1 must include the service independent configuration of both the SAR and the CPCS and the configuration of any parameters that are determined by the service itself (see Figure 7.7). AAL1 generates an alarm if it receives no cells for an extended period of time, and this period of time is configurable. For circuit emulation, there is also configuration of the size of the CDV buffer and of the way that channel associated signaling in telecommunications is carried.

The measures that are used for performance monitoring cover problems with the SAR header and with cell sequence numbering, problems with synchronization in the CPCS, and problems with the buffer used in circuit

Parameter = Attribute		Notes
AAL 1 Profile	Bit Rate	Integer value
	Synchronization — Clock	sync/SRTS/adaptive
	Synchronization — Frame	SDT (yes or no)
	Data subtype	Unspecified, video, audio (voice/high quality) circuit emulation (sync/async)
	FEC method	none/delay sensitive/loss sensitive
	Partially filled cells	number of leading octets used
CE Profile	Alarm	Intergration perion for loss cells
	CDV buffer	Size (in units of 10 microseconds)
	CA signalling	E1/DS1-SF/DS1-ESF/J2

Figure 7.7　AAL1 configuration parameters (represented as attributes).

emulation (see Figure 7.8). The measures of AAL1 performance are carried in current data objects and history data objects, while sets of configuration parameters are held in AAL1 profiles (see Figure 7.3).

For example, to change the size of the CDV buffer for an existing circuit emulation profile, the OS writes the new buffer size in the attribute in the profile object. To obtain the number of AAL1 SAR header errors in the current monitoring period, the OS reads the header errors attribute in the AAL1 current data object contained in the VC TTP object that represents the end-point of the VCC.

7.2.5　Synchronous Variable Rate Traffic (AAL2)

AAL2 is not popular because the services that it was designed for lack the simplicity of constant rate services and the delay insensitivity of data communications services. In addition, the variable rate codecs that could use AAL2 are not common and are more difficult to use. It is more sensible to include rate smoothing buffers in a variable rate codec since there is a greater market for constant rate codecs, and these can be used over AAL1. In practice AAL2 is not used.

Catagory	Measure = Attribute
SAR header	Header errors
SAR sequence numbers	Cell loss Cell misinsertion Sequence violations
Synchronization (CPCS)	SDT pointer parity failures SDT reframes
Buffer (circuit emulation)	Underlows Overflows

Figure 7.8 AAL1 measures of performance (represented as attributes).

7.3 Asynchronous Traffic—Original (AAL3/4)

Initially, AAL3 was defined for connection-oriented data communications and AAL4 was defined for connectionless data communications. These were subsequently combined into AAL3/4 because they could use the same SAR and CPCS since the differences between them were specific to their client services.

Both the SAR sublayer and the CPCS include a number of header and trailer fields whose value is questionable. In particular, the SAR sublayer includes a CRC that should really belong to the CPCS, which can also perform it more efficiently. The SAR sublayer also includes client multiplexing, which should really belong to the SSCS.

7.3.1 AAL3/4 Segmentation and Reassembly

AAL3/4 has a more complex SAR sublayer than AAL1 because it includes error detection for the SAR payloads. The SAR header consists of a two bit Segment Type (ST) field, a four bit Sequence Number (SN), and a ten bit Multiplex Identification (MID) (see Figure 7.9). The SAR trailer consists of a six-bit Length Indicator (LI) field and a ten-bit CRC.

The ST indicates if the SAR payload is a Single Segment Message (SSM) or if it is the beginning, continuation or end of a multi-segment message (BOM, COM, or EOM). The SN allows missing segments of a multi-segment message to be identified.

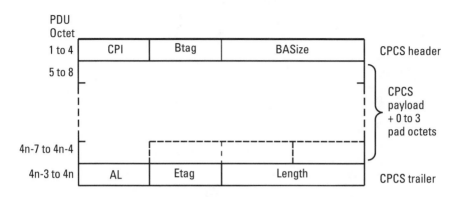

Figure 7.9 The structure of ATM cells used in AAL3/4.

The MID marks a clear difference between AAL3/4 and AAL5. It allows AAL3/4 to carry multiple client services on the same VCC and makes AAL3/4 consistent with the IEEE 802.6 protocol. The clients of AAL5 must agree on their own multiplexing if they wish to share a VCC since AAL5 has no equivalent label, but more often than not each client service has a dedicated VCC.

The LI should have the value 44 unless the payload is an SSM or an EOM. In this case the value should be a multiple of four because zero to three octets of padding will have been added. A value of 63 indicates that a multi-segment message has been aborted.

7.3.2 AAL3/4 Convergence Sublayer (CS)

The AAL3/4 CPCS maps its Protocol Data Units (PDUs) onto segments that are carried in the SAR payloads. If a CPCS PDU is less than 44 octets long then it is carried as an SSM. If the PDU is more than 44 octets long then it is carried as a multisegment message.

Each complete CPCS PDU is a multiple of four octets long and has a header and a trailer (see Figure 7.10). The Common Part Indicator (CPI) field in the header was defined with the intention of specifying the units for the Length and Buffer Allocation Size (BASize) fields. The only agreed value for the CPI field is all zeroes, and this indicates that the units are octets.

The BASize indicates the size of buffer needed to reassemble the message. This can be greater than the length of the PDU if the data is streamed. The Length field indicates the length of the CPCS payload and so that any pad octets can be removed.

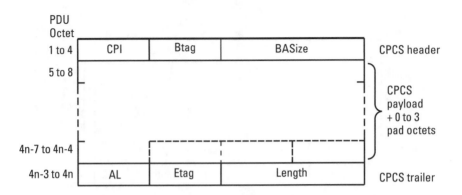

Figure 7.10 The structure of the CPCS Protocol Data Unit (PDU) for AAL3/4.

The Beginning Tag (Btag) in the header has the same value as the Ending Tag (Etag) in the trailer. These two tags confirm that the header and trailer belong to the same CPCS PDU. The Alignment (AL) field is merely additional padding that makes the trailer four octets long.

7.3.3 AAL3/4 Management and Modeling

The configuration for AAL3/4 is simpler than that for AAL1. Only the range of MIDs need to be configured for the SAR sublayer. The CPCS needs configuration of the maximum length of its PDUs, but the header and trailer tags do not need to be configured because they are handled automatically. The CPCS also needs configuration of the mode of operation and the type of communication. Sets of configurable parameters are held in AAL3/4 profiles (see Figure 7.11).

The SAR sublayer can detect cells with unexpected SNs and MIDs and it can detect cells with invalid LIs and incorrect CRCs. It can also count the number of multi-segment messages that have been aborted due to a value of 63 in the LI field or due to reassembly timing out.

The CPCS can detect if the CPI or AL fields are not zero. It can also detect if the Length field is inconsistent with the length detected by the SAR sublayer and if there are inconsistencies between the message length and the BaSize, or between the Btag and the Etag.

The AAL3/4 measures of performance (see Figure 7.12) are recorded in current and history data objects (see Figure 7.3).

For example, to define a new AAL3/4 profile with a higher limit on the MID identifier, the OS creates a new profile object which has the new limit

	Parameter = Attribute	Notes
SAR	Multiplex ID Range (AAL 3/4 only)	Lowest to highest
Convergence sublayer	Maximum CPCS PDU size	Forward and backwards
	AAL Mode	Message/streaming Assured/non-assured
	SSCS Type	Unspecified/ data (assured/non-assured)/ frame relay

Figure 7.11 Configuration of ATM adaptation for data communication (AAL3/4 and AAL5).

	Measure = Attribute	Notes
SAR	Sum of invalid fields	Multiplex ID Length field
	Unexpected SN	On COM or EOM cells
	Unexpected MID	Mismatched BOM and EOM cells
	CRC violations	Total number
	Sum of incorrect fields	Unexpected sequence number Unexpected Multiplex ID CRC violation
	Timeouts	On reassembly
	Aborts	Indicated in EOM cell
CPCS	Sum of invalid fields	Common Part Indicator (CPI) Buffer allocation size Alignment
	Tag mismatch	Beginning vs ending
	Buffer length mismatch	Buffer size vs length field
	PDU length mismatch	PDU size vs length field
	Sum of incorrect fields	Tag mismatch Buffer length mismatch PDU length mismatch

Figure 7.12 AAL3/4 measures of performance (represented as attributes).

in the high field of the MID range attribute. To obtain the number of tag mismatches in the last monitoring period, the OS reads the value of the tag mismatch counter of the history data object contained in the AAL3/4 current data object for the VCC end-point.

7.4 Asynchronous Traffic—Streamlined (AAL5)

AAL5 was devised as a more efficient alternative to AAL3/4. The differences between connection-oriented and connectionless operation are service specific and in AAL5 these are assigned to the SSCS (see Figure 7.1). The service independent CPCS performs the integrity checking that follows the reassembly of PDUs, allowing the AAL5 SAR sublayer to be kept to a minimum and simplifying the task because it is not necessary to check each cell.

7.4.1 AAL5 Segmentation and Reassembly

The AAL5 SAR sublayer has no headers or trailers and its payload is the complete payload of the ATM cell. Only AAL5 uses a bit in the ATM cell headers, the AAL indication bit which is the least significant bit of the PT field (see Figure 7.13). The SAR sublayer indicates that an ATM cell carries the end of a PDU by setting this bit to one. The next ATM cell is the first ATM cell of the next PDU.

The SAR sublayer, like the intermediate nodes of the VCC, can indicate congestion and request a reduction in the data rate by setting the

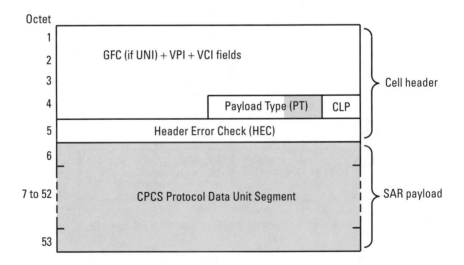

Figure 7.13 The structure of ATM cells used in AAL5.

Explicit Forward Congestion Indication (EFCI) bit in cell headers. This bit is the middle bit of the PT field. The most significant bit of the PT field must be zero to indicate that the cell carries user traffic. The SAR sublayer can also set the CLP bit in cell headers to mark cells that should be deleted first.

7.4.2 The Common Part of the AAL5 Convergence Sublayer

The CPCS of AAL5, like its SAR sublayer, is much simpler than that of AAL3/4. A CPCS PDU is carried as a sequence of ATM cell payloads. The PDU has a trailer, but no header (see Figure 7.14).

The length field is included because pad octets are added to fill the last cell of the PDU. The CPCS-UU octet carries user-to-user information transparently between the clients of the CPCS. The CPI octet is not used and its default value is zero.

The final field in the PDU trailer is a 32-bit CRC. There is no sequence numbering in the SAR sublayer because the CRC detects most errors. The probability that alarms have not been generated before the error rate becomes high enough to create errors that the CRC cannot detect is less that than the probability of other hardware or software faults affecting the client service. The Length field also assists here by indicating if cells are missing or if additional cells have been inserted.

Neither the CPCS of AAL5 nor its SAR sublayer support multiplexing of client services. If multiplexing is needed then it must be handled either by the SSCS or by the client services themselves. This is more appropriate than

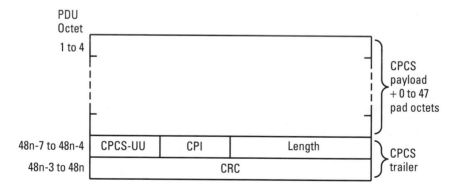

Figure 7.14 The CPCS Protocol Data Unit (PDU) for AAL5.

the AAL3/4 approach because multiplexing is service specific and many client services do not need it.

Some client services do not even need an SSCS. IP is carried directly in AAL5 PDUs (see Section 7.4.4) This makes IP applications over ATM easier to implement than conventional UNI or NNI signaling because IP is more widely deployed.

7.4.3   AAL5 Management and Modeling

The configuration of AAL5 is even simpler than the configuration of AAL3/4 (see Figure 7.11) because the SAR sublayer does not contain MIDs. Only the CPCS needs to be configured. As in AAL3/4, for the CPCS it is necessary to specify the maximum size of the PDU, the mode of operation and the type of client service.

Performance monitoring for AAL5 is also simpler than for AAL3/4. There are no SAR fields to monitor but the SAR sublayer can note how often the reassembly of PDUs times out. Once the PDUs are reassembled they may be discarded if they are too long, or if there is a CRC failure, or if their length is inconsistent with the Length field, or if the CPI field is invalid (see Figure 7.15).

Profile objects for AAL5 only have three of the four attributes of the profile objects for AAL3/4 (see Figure 7.11) because AAL5 has no multiplex identifiers. The current data objects for AAL5 are simpler and cleaner than those for AAL3/4 (compare Figures 7.12 and 7.15).

For example, to define a new AAL5 profile with a lower limit on maximum backward CPCS PDU size, the OS creates a new AAL5 profile object with the new lower limit in the backwards field of the CPCS PDU maximum size attribute. To discover the number of CRC violations in the current

	Measure = Attribute	Notes
SAR	Timeouts	On reassembly
CPCS	CRC violations	Total Number
	Sum of invalid fields	Common Part Indicator (CPI) OVersized Length inconsistency

Figure 7.15   AAL5 measures of performance (represented as attributes).

monitoring period, the OS reads the value of the CRC violations counter contained in the VCC TTP object that represents the end-point of the VCC.

7.4.4 Internet Protocol (IP) over AAL5

It was agreed to fragment IP datagrams that are longer than 9180 octets if they carried over ATM so as to be consistent with the Switched Multi-megabit Data Service (SMDS). Although this is an unnecessary restriction, it is no more of a problem than for SMDS and software developed for SMDS could be re-used.

The fragmentation and reassembly of IP datagrams has similarities to the segmentation and reassembly of AAL5 PDUs. Each fragment has a header that is analogous to the header of an ATM cell and contains a flag bit that is similar to AAL indication bit in an ATM cell header, although the polarity of the bit is opposite

The transport of IP fragments differs from the transport of ATM cells because the fragments are not guaranteed to arrive in the same order that they were sent. The value of the fragment offset in their headers indicates the location of the fragment payload in the reassembled IP datagram. Although ATM guarantees that fragments that it creates will arrive over an ATM connection in the same order that they were sent, normal IP fragmentation is used because the destination of the fragments may be outside the ATM network.

Nothing more is needed if IP is the only protocol that is carried across the ATM VCC. If IP shares the VCC with other protocols then a prefix is added to IP datagrams to identify them as IP. This prefix consists of the IEEE 802.2 Logical Link Control (LLC) field followed by a SubNetwork Attachment Point (SNAP) field. The coding of these fields is fixed for IP (see Figure 7.16) and the SNAP Type subfield has the same value as when IP is multiplexed over Ethernet. Whether or not the LLC/SNAP prefix is added is a matter for the client services, but the spread of IP to carry application protocols increasingly makes it an unnecessary overhead.

7.5 The Signaling ATM Adaptation Layer (SAAL)

The SAAL is used by ATM call control signaling. It is the combination of AAL5 and the SSCS for ATM signaling (see Figure 7.1). The function of the SSCS is to map the messages of the ATM call control signaling to and from the PDUs of the AAL5 CPCS.

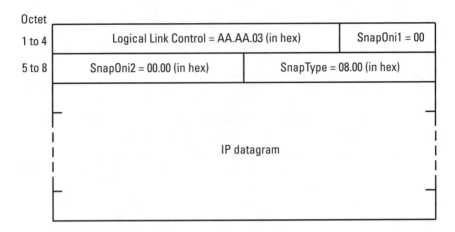

Octet

| 1 to 4 | Logical Link Control = AA.AA.03 (in hex) | SnapOni1 = 00 |
| 5 to 8 | SnapOni2 = 00.00 (in hex) | SnapType = 08.00 (in hex) |

IP datagram

Figure 7.16 The IP prefix for multiplexed protocols.

The SAAL is used for signaling on both User-Network Interfaces (UNIs) and Network-Node Interfaces (NNIs). It is also used for signaling on VB5 interfaces, which connect access networks to their host service nodes.

The SSCOP provides both assured and non-assured communications to the different SSCFs. The assured communication is similar to that provided by the IP Transmission Control Protocol (TCP). It provides assurance that messages are delivered without corruption and in the correct order. The non-assured communication is similar to that provided by the IP User Datagram Protocol (UDP) and the loss of data sent in this way may not be detected.

ATM call control on the UNI and on the NNI use different SSCFs. Although the VB5 interface (see Chapters 15–17) is a specialized NNI, it uses the SSCF for the UNI so that all access signaling uses the same SSCF.

7.5.1   The Service-Specific Connection-Oriented Protocol (SSCOP)

The different types of SSCOP PDUs are identified by the value of their type field. This field is the only one that is common to all PDUs (see Figure 7.17).

Only two types of PDU are defined for non-assured communication, which has no means of detecting lost PDUs. One of these is for payload data and the other is for management overheads (see Figure 7.18). The only other fields in these PDUs are the payload information field and the pad length field that indicates how may padding octets have been added to give a multiple of four.

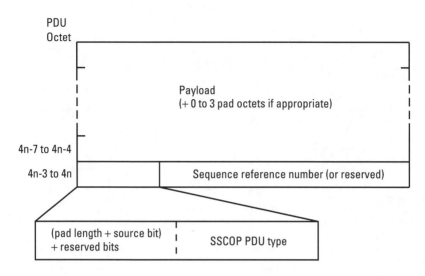

Figure 7.17 The structure of SSCOP PDUs.

Several types of PDU are defined for assured communication so that information can be delivered without corruption and in the correct order. Some PDUs are used to confirm that the information has been delivered, but more are used to control the logical connections over which the information flows. The presence or absence of a logical connection indicates whether or not assured communication is possible.

Many of the PDUs for assured communication include sequence numbers. These can identify the PDUs that contain them or refer to other PDUs. There are three type of sequence number, one that labels the client data, one that refers to client data, and one that labels the messages that control the logical connections that carry the client data.

Assured communication relies on the CRC in AAL5 CPCS to discard PDUs that have been corrupted. The SSCOP assumes that those PDUs it receives are not corrupted. For this assumption to be valid, detected errors must cause the closure of logical connections before the probability of an undetected error becomes non-negligible. The SSCOP uses a sophisticated approach to the detection and retransmission of lost PDUs that reduces unnecessary retransmissions and controls the flow of further data.

7.5.1.1 Delivery Confirmation: Status and Poll Messages

Client information for assured delivery is carried in sequenced data PDUs. Each of these is labeled with a data sequence number (see Figure 7.18) so

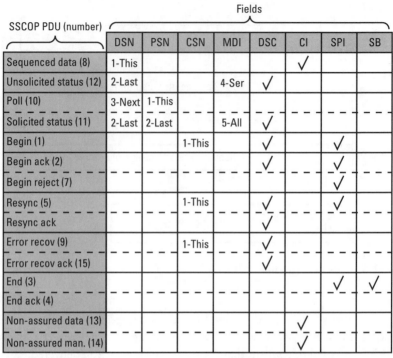

SSCOP PDU (number)	DSN	PSN	CSN	MDI	DSC	CI	SPI	SB
Sequenced data (8)	1-This					✓		
Unsolicited status (12)	2-Last			4-Ser	✓			
Poll (10)	3-Next	1-This						
Solicited status (11)	2-Last	2-Last		5-All	✓			
Begin (1)			1-This		✓		✓	
Begin ack (2)					✓		✓	
Begin reject (7)							✓	
Resync (5)			1-This		✓		✓	
Resync ack					✓			
Error recov (9)			1-This		✓			
Error recov ack (15)					✓			
End (3)							✓	✓
End ack (4)								
Non-assured data (13)						✓		
Non-assured man. (14)						✓		

DSN = Data Sequence Number
PSN = Poll Sequence Number
CSN = Connection Sequence Number
MDI = Missing Data Identification
DSC = Data Sequence Credit
CI = Client Information (+ pad + pad length)
SPI = SSCOP Peer Information (+pad + pad length)
SB = Source Bit

1-This = sequence number for this message
2-Last = sequence number of last received message
3-Next = data sequence number of message to be sent next
4-Ser = two data sequence messages that identify a single series
          of missing data
5-All   = sequence numbers that identify all missing data

Figure 7.18 SSCOP messages and fields.

that PDUs that have already been received can be ignored and PDUs that are lost can be identified.

When the sequence numbering indicates that PDUs have been lost, an unsolicited status message is returned. This specifies a sequence of missing PDUs (see Figure 7.19). The number of the first PDU that has not been received may differ from the first number of the newly identified missing series if other status messages have already been sent.

Status messages may have already been sent either because the transmitter has solicited them or because the receiver has detected a missing sequence. A solicited status message contains more information than an unsolicited

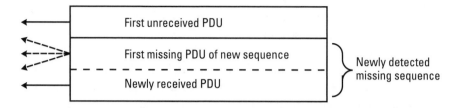

Figure 7.19 Data sequence numbers in unsolicited status messages.

message because the transmitter does not know about previous unsolicited messages that have been lost. The transmitter sends a poll message to solicit a status message. Poll messages have their own sequence numbers and they tell the receiver the number of the first sequence data PDU that has not been sent. This allows the receiver to return a solicited status message that identifies all PDUs that have not been received (see Figure 7.20).

To allow responses to be matched to requests, each solicited status message returns the poll sequence number of the poll message that solicited it. If no data is missing then the solicited message returns the number of the first unreceived data sequence number. Solicited status messages identify the start of each gap by the first number that was not received. If the gap has a known end then the status message includes the next number that was received.

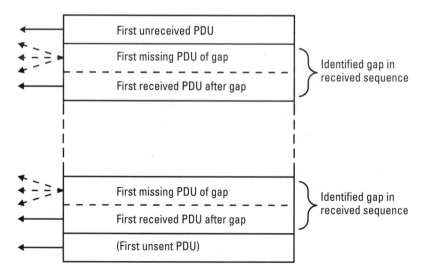

Figure 7.20 Identification of missing PDUs in solicited status messages.

It would be better if unsolicited status messages indicated all of the missing information in case previous messages were lost. It is also redundant for status messages to identify the first unreceived PDU and to include it again when identifying missing information, and there is no need for solicited status messages to return the number of the first unsent PDU because this is implied by the returned poll sequence number.

7.5.1.2   The Control of Logical SSCOP Connections

The presence of a logical connection indicates that assured communication is possible. A connection is initiated by sending a begin request. The reply to this is either an acknowledgement, if the other side agrees to establish the connection, or a rejection. Once the connection is established, assured communication can take place until an end request or an end acknowledgement terminates the connection.

Each begin message contains a connection sequence number (see Figure 7.18 and 7.21). Error recovery requests and resynchronization requests are also identified by their connection sequence numbers. Like begin requests, these also cause a reset of the data sequence numbering, but resynchronization requests can also carry information between peer SSCOP signaling entities. There are no rejection messages for error recovery or resynchronization because end messages are used to terminate the logical connection if these requests cannot be accepted.

End requests contain a flag that indicates whether it is the SSCOP client or the SSCOP itself that has requested the termination of the logical connection.

7.5.1.3   Flow Control by Credit Allocation

Each side of a logical connection controls the flow of assured information by allocating credit to the other side. The side that receives the credit allocation can send sequenced data PDUs with numbers up to the credit limit.

Credit is allocated initially in begin requests and acknowledgements. Once the connection is established, the credit allocation is updated by status messages. Credit can also be allocated by resynchronization and error recovery messages that keep the logical connection open when problems have occurred.

7.5.2   The Service Specific Coordination Functions (SSCFs) for the SAAL

The assured and non-assured communications provided by the SSCOP are used by one type of SSCF for UNI signaling and by another type of SSCF for

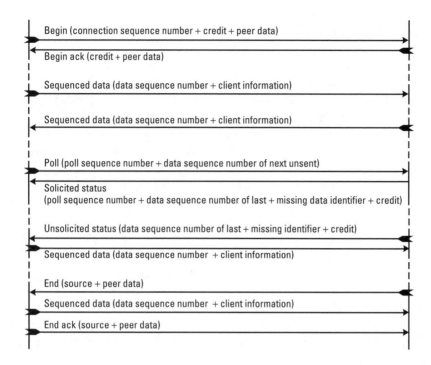

Figure 7.21 Example of assured communication.

NNI signaling. Both types of SSCF support point-to-point signaling, but only the SSCF for the UNI supports broadcast signaling.

Broadcast signaling is not assured. This differs from TCP because IP datagrams specify their own destinations. The logical connection specifies the end-points for assured SSCOP communication, and this connection is between two points.

The SSCF for the UNI supports both assured and non-assured communications. Non-assured communication for the UNI can in theory be used for both point-to-point and broadcast signaling. In practice, the SSCFs for both the UNI and the NNI use assured communication for point-to-point signaling. The SSCF for the NNI provides greater visibility and control of the SSCOP.

7.5.3   Management Modeling of the SAAL

The management modeling of the SSCOP, like that of the ATM adaptation layers, uses profile objects that other objects refer to. This is described in more detail in Chapter 8 because additional access objects are also needed.

7.6   Comments on ATM Adaptation

Perhaps the only merit of AAL2 is that, by illustrating the difference between concepts and reality, it provides the basis of a lesson in Zen Buddhism. This may be unfair because AAL2 also has the merit that it can be ignored. Unfortunately the SAAL is not as easy to ignore.

In the light of the widespread use of ATM for the transport of IP, it is sensible to ask why the SAAL is needed for ATM signaling. ATM call control could be redefined as an IP application, using TCP for assured communication and UDP for non-assured communication. This would help IP routers to set up ATM connections and it could also facilitate the introduction of ATM call control.

A further step could also be taken because AAL5 can be further simplified for the transport of IP. This is because all that the AAL needs to do is to use the AAL indication bit to mark the start of IP datagrams since datagrams indicate their own length and IP application protocols perform their own integrity checking.

For the existing AAL5, the monitoring and control by the AAL5 SAR sublayer of bits in the payload type field in the ATM cell headers is not visible to the management systems. Normally changes in cell loss priority are monitored at the adaptation functions between the ATM layers because this is where the changes occur. Any changes in the cell loss priority or in the congestion indication that are made by the AAL5 SAR sublayer are handled automatically and are invisible to the management systems.

7.7   Summary

The ATM Adaptation Layer (AAL) matches the client services carried by ATM to the payloads carried by ATM cells. Five AALs were originally defined. AAL1 is the preferred AAL for services like real-time audio and video that need a synchronization signal. AAL5 is the preferred AAL for data communications services and IP applications that do not require a synchronization signal.

AAL2 is largely ignored because it was defined for the exotic, variable transmission rate form of the services that AAL1 supports. Layers three and four were combined into AAL3/4 but this is less common than the simpler AAL5.

Each AAL contains a segmentation and reassembly (SAR) sublayer that breaks client traffic down into the payloads of ATM cells and rebuilds the client traffic from the ATM cells. The common part of the convergence sublayer (CPCS) sits above the SAR sublayer and confirms its operation.

The synchronous residual time stamp (SRTS) method allows the AAL1 SAR to generate a clock for the client service, and the structured data transfer (SDT) method allows it to generate a framing signal. Circuit emulation over AAL1 emulates constant rate physical transmission and requires a buffer to compensate for cell delay variation (CDV) on the ATM connection.

The SAR sublayer of AAL5 is simpler than that of AAL3/4, partly because AAL5 does not check each cell separately. AAL5 also does not support client multiplexing because this is a service specific function.

A service specific convergence sublayer (SSCS) is defined for ATM call control signaling. This sits above AAL5. Its lower part, the Service Specific Connection-Oriented Protocol (SSCOP) is common to signaling on both the user-network interface (UNI) and the network-node interface (NNI). The upper part, the service specific coordination function (SSCF), has one form for the UNI and another form for the NNI. The combination of AAL5 and the SSCS is called the signaling ATM adaptation layer (SAAL). Assured communication over the common SSCOP corresponds to the Transmission Control Protocol (TCP) for the Internet Protocol (IP) and nonassured communication corresponds to the User Datagram Protocol (UDP).

The management model for ATM adaptation uses profile objects to represent sets of parameters for AALs. Groups of VCC TTP objects, which represent the end-points of virtual channel connections (VCCs), point to their common profile. The management model also uses current data objects to hold the results of ongoing performance monitoring of AALs. These are contained within the VCC TTP objects and they in turn contain history data objects that hold the records of previous monitoring.

References

[1]   McDysan, D. E., and D. L. Spohn, *ATM Theory and Application*, New York: McGraw-Hill, 1995.

[2]   *B-ISDN ATM Adaptation Layer (AAL), Type 5 Specification*, ITU-T Recommendation I.363.5, 1996.

[3]   Comer, D. E., *Internetworking with TCP/IP: Volume 1 - Principles, Protocols and Architecture*, New Jersey: Prentice-Hall, 1995.

[4]   *B-ISDN ATM Adaptation Layer - Service Specific Connection Oriented Protocol (SSCOP)*, ITU-T Recommendation Q.2110, 1994.

[5]   *B-ISDN ATM Adaptation Layer - Service Specific Co-ordination Function for the Support of Signaling at the User Network Interface (SSFC at UNI)*, ITU-T Recommendation Q.2130, 1994.

[6]   *B-ISDN ATM Adaptation Layer - Service Specific Co-ordination Function for the Support of Signaling at the Network Node Interface (SSFC at NNI)*, ITU-T Recommendation Q.2140, 1995.

[7]   *B-ISDN Signaling ATM Adaptation Layer (SAAL) overview description*, ITU-T Recommendation Q.2100, 1994.

[8]   *Broadband Switch Management*, ITU-T Recommendation Q.824.6, 1998.

[9]   Clark, G. C. Jr., and Cain, J. B., *Error-Correction Coding for Digital Communications*, New York: Plenum Press, 1981.

8

ATM Signaling

Tweedledum and Tweedledee agreed to have a battle...
— Lewis Carroll, *Through the Looking-glass*

ATM signaling is more complex than narrowband signaling because it handles a wider range of connections and services. With hindsight, ATM signaling should have been made simpler, more robust, and more easily extendible because this would have been less restrictive and led to faster implementations.

On the other hand, the world would be a less interesting place if a more effective approach had been taken to ATM signaling. This is because we could have missed the recurrent battles between ITU-T and the ATM Forum (ATMF), and the specter of the Internet Protocol being used in the future for ATM signaling.

8.1 Background

ATM connections are more flexible than narrowband connections, and this is reflected in ATM signaling. The connections can have a wider range of data rates, and the data rate can be modified after the connection is established. Connections are also classified, e.g. Constant Bit Rate (CBR) and Variable Bit Rate (VBR), according to the parameters that apply to them.

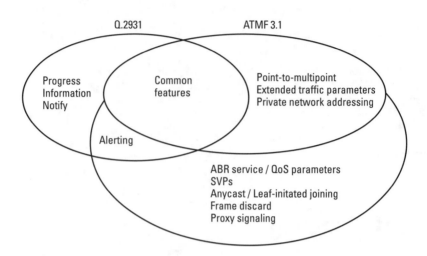

Figure 8.1 Features of UNI signaling.

VBR traffic has a Peak Cell Rate (PCR), a Sustainable Cell Rate (SCR), and a Maximum Burst Size (MBS). CBR traffic, in contrast, only has a PCR since its rate is constant by definition. A connection also has Quality of Service (QoS) parameters that can specify its Cell Loss Ratios (CLRs), its Cell Transfer Delay (CTD) and its Cell Delay Variation (CDV).

8.1.1   ITU-T and the ATM Forum

The ITU-T originally decided to organize its work according to Capability Sets (CSs). Capability Set 1 (CS-1) supports basic point-to-point calls and traditional supplementary services, such as calling line identification. CS-2 supports additional transfer capabilities, such as best effort, and the modification of connection parameters during a call. CS-3 supports more sophisticated connection topologies with more than two parties. It was soon recognized that this classification did not allow standards to be produced quickly enough, and a more pragmatic approach had to be adopted.

ATM signaling requirements are different at different points in a network. ITU-T has produced Q.2931 [1] Digital Subscriber Signaling System No. 2 for signaling on the user-network interface (UNI) and Q.2761 to Q.2764 [2] for the network-network interface (NNI).

The ATMF decided to base its signaling on the ITU-T UNI, but to modify it to take account of what equipment suppliers thought was practical.

This led the ATMF to develop its own specifications for the UNI, in particular Versions 3.1 and 4.0 [3] (see Figure 8.1). These cover all aspects of the interface from the physical layer, through the ATM and adaptation layers, right up to the signaling and the services that it supports. There is an earlier version of the ATMF UNI specification, Version 3.0 that is not compatible with Q.2931 because it uses a different signaling ATM adaptation layer (SAAL). The ATMF also developed a symmetrical version of the signaling for the NNI within a single network [4] (the intra-NNI) and for the NNI between different networks [5] (the inter NNI).

8.1.2 Interworking

A major difference between ITU-T and the ATMF is the way ATM signaling relates to physical interfaces. In the ATMF, the signaling refers to ATM Virtual Path Connections (VPCs) on a specific physical interface. In the ITU-T a more general approach is used that allows VPCs to be decoupled from physical interfaces, since this supports cross-connection between the signaling end-points. Since a Virtual Path Identifier (VPI) is specific to a physical interface, the ITU-T signaling uses VPC Identifiers (VPCIs) that are independent of physical interfaces but that map onto the VPIs at a physical interface. Although the ITU-T approach is more general, it requires co-ordination of the various mappings to the VPIs.

Interworking between B-ISUP and UNI 3.1 signaling is difficult because UNI 3.1 was designed for terminals that answer calls automatically and do not use the ALERTING message that corresponds to the indication of remote ringing in PSTN. Although UNI 4.0 includes an ALERTING message, interworking with B-ISUP remains difficult because B-ISUP has lagged behind UNI 4.0 in its support of more advanced features.

8.2 Services, Addresses, and Topology

8.2.1 Bearer Services and Telecommunications Services (Teleservices)

ATM services (see Figure 8.2) can be classified as bearer services or telecommunications services (teleservices). Pure bearer services correspond to connection types, such as CBR and VBR, but bearer services can also include the ATM Adaptation Layer (AAL) that customizes the connection to a particular application. Teleservices, such as teleconferencing, also include a signaling application that operates above the AAL.

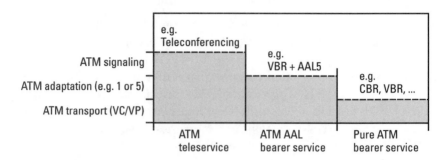

Figure 8.2 ATM bearer services and teleservices.

Unfortunately, it can take longer to standardize a teleservice that it takes to develop it, and this led to the concept of semistandardized teleservices. In these the user can customize the service by selecting some attribute, such as its bandwidth, from a specified list. Alternatively, the user can customize the service by selecting a particular feature for a particular call, such as audio only for a video-telephony call. Unfortunately defining semistandardized teleservices adds further delays because more work is needed to specify the list of options.

A better approach is the user-defined teleservice. In this case user defines the service in terms of the network resources that are needed for a call. The network operator is responsible for ensuring that the necessary resources are provided, but not for ensuring the compatibility of the user equipment.

An open teleservice is an attempt to have the best of all worlds. The service is open in the sense that it can be a combination of standardized and non-standardized components. These components are only defined when the call is set up and can be modified during the lifetime of the call. An open teleservice does not confirm to the KISS (Keep It Simple, Stupid) principle. Supplementary services (see Figure 8.3) add additional features to bearer services and teleservices. These features are often a subscription option that can be applied to all calls or to specific calls.

8.2.2 ATM Addresses

There are two types of ATM addresses. The numbering scheme used for narrowband telephony led to E.164 numbers that are favored by ITU-T. The ATMF prefers to use AESAs (ATM end-system addresses) that arose from data communications.

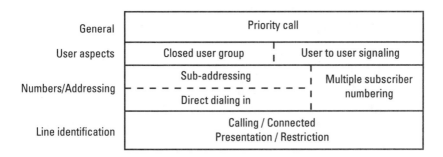

General	Priority call	
User aspects	Closed user group	User to user signaling
Numbers/Addressing	Sub-addressing	Multiple subscriber numbering
	Direct dialing in	
Line identification	Calling / Connected Presentation / Restriction	

Figure 8.3 ATM Supplementary Services.

The 15 or less E.164 digits consist of a subscriber number, an optional national destination code that specifies a domain within a country, and an optional country code. E.164 numbers can be incorporated into AESAs.

AESAs are 20 octets long. They begin with an authority and format identifier (AFI), which indicates the structure of the AESA. The AFI indicates whether the AESA contains an E.164 number or a domain specific part (DSP) that is registered with a national body or an international organization.

8.2.3   Connection Topologies

A simple connection joins only two parties. In general a connection joins a number of parties. Information can flow between any pair of parties in three ways, namely in either direction or in both directions.

Basic ATM signaling only supports two connection topologies. The first of these is the simple case of two parties, known as a point-to-point connection. The second case is uni-directional connection from one party (the root) to all other parties (the branches). This type of connection is known as a point-to-multipoint connection. Point-to-multipoint connections are feasible because ATM switches can replicate ATM cells, allowing the root to talk to the branches. The branches were not allowed to talk to the root because of unresolved problems if cell flows merge.

These two cases are easy to handle because connections are normally initiated by one of the parties in the connection, the calling party. For point-to-point connections the calling party is one of the two ends. For point-to-multipoint connections the root is the calling party.

It is also possible that a third party that is not involved in the connection has set it up. Third-party signaling is interesting because it opens the door to alternative approaches to connection control.

8.3 UNI Signaling

The ITU-T and ATMF specifications for the UNI have many parts in common. Certain of the messages in Q.2931 are not supported by the ATMF UNI specifications (see Figure 8.1). However the ATMF specifications support a wider range of features.

These differences present a problem to network operators, because networks are simpler if they do not support both ITU-T and ATMF specifications. The ATMF specifications are often preferred because they are simpler and have greater capabilities, but in practice the true casualty has been the introduction of UNI signaling, since it is even simpler for network operators to delay making a choice.

8.3.1 ITU-T UNI Signaling

The three phases of an ATM call are establishment, in-call control, and release. Call establishment (see Figure 8.4) is similar to that for Narrowband ISDN (N-ISDN).

Either party can send a RELEASE message to terminate a call, and a RELEASE COMPLETE message is returned in response (see Figure 8.12). Q.2931 has a two-step procedure for call release, unlike N-ISDN signaling (Q.931) that has three steps.

During the call STATUS messages are sent if an anomalous condition has been detected or in response to a STATUS ENQUIRY message. PROGRESS, SETUP ACKNOWLEDGE and INFORMATION messages are defined to support interworking with N-ISDN, and a NOTIFY message, as in N-ISDN provides information about supplementary services.

The UNI signaling also supports point-to-multipoint operation (see Figure 8.5). There is an ADD PARTY request and corresponding rejection or acknowledgement messages, and a DROP PARTY message that can only be acknowledged. There is also a PARTY ALERTING message that like the ALERTING message is excluded from the ATM Forum's UNI v3.1 specification.

The MODIFY message, with its associated rejection and acknowledgement messages, allows connection parameters to be negotiated after a call has been established. The CONNECTION AVAILABLE message can be requested as positive confirmation of a modification.

Q.2931 also supports a look-ahead procedure that allows the calling party to confirm that the required service can be supported. In particular the look-ahead procedure can confirm that the called terminal is compatible and

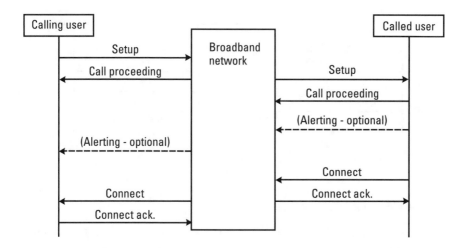

Figure 8.4 Simple ATM call establishment at the UNIs.

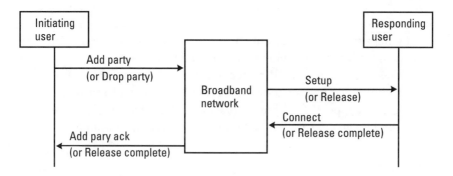

Figure 8.5 In-Call Control of Multipoint Connections.

free. The SETUP message can be used to invoke a look-ahead and the CALL PROCEEDING message carries the response. Look-ahead causes the network to exchange FACILITY messages with the called party (see Figure 8.6). The network then decides to proceed by sending a SETUP message to the called party or to reject the call by sending a RELEASE message to the calling party.

In the event of a serious problem, RESTART messages are sent to request the release of all calls controlled by a particular signaling channel, and associated acknowledgements are returned.

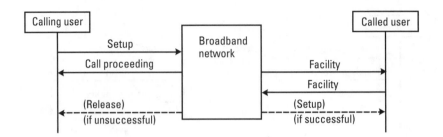

Figure 8.6 Look-ahead.

8.3.2 ATM Forum UNI Signaling

Version 4.0 of the ATMF UNI specification adds additional features to those in Version 3.1. It supports available bit rate (ABR) connections and independent Quality of Service (QoS) parameters (see Chapter 19 for a description of connection types and parameters). It supports switched VPCs and extends the multicasting supported by Version 3.1 to allows branches to control connections.

Perhaps the most significant change in Version 4.0 is the introduction of proxy signaling, which allows connections to be established by a third party. This could allow an IP router to set up a connection between two other machines that had no knowledge of ATM signaling. In theory, ITU-T signaling also supports this, but in practice ITU-T have lagged behind in separating the signaling end-points from the connection end-points.

8.4 NNI Signaling

The distinction between inter-NNI signaling and intra-NNI signaling arises from the architectural model used by the ATMF. In this model, there are both public and private networks and these can have internal NNIs (intra-NNIs) or external NNIs (inter-NNIs) between their switches (see Figure 8.7). Public networks use the Broadband ISDN User Part (B-ISUP) of ITU-T's Signaling System No. 7 (SS7) for intra-NNI signaling, whereas private networks use the Private Network-Network Interface (PNNI).

In the ITU-T view, B-ISUP is both used within networks and between different networks. The ATMF took the view that a different interface was needed for carriers between public networks and defined the Broadband Inter-Carrier Interface (B-ICI) as the inter-NNI. Although this mirrors the

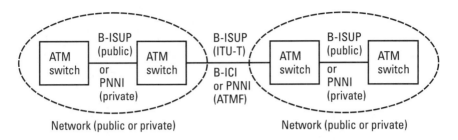

Figure 8.7 Network-Network Interfaces.

difference between internal and external routing in IP networks, the distinction is artificial in ATM networks.

This has lead to a curious twist in this story. There were delays in the finalization of both B-ISUP and B-ICI, partly due to their complexity and partly due to loss of momentum once PNNI was defined. The implementation of PNNI, coupled with these delays, has undermined the adoption of B-ISUP and B-ICI, and PNNI is often preferred for both the intra-NNI and the inter-NNI.

8.4.1 ITU-T NNI Signaling

The B-ISUP on SS7 is the ITU-T standard for NNI signaling. Like N-ISUP, B-ISUP uses common channel signaling, which has a dedicated signaling channel that carries the signaling for a number of bearer connections.

Signaling messages are carried in a Virtual Channel (VC), often that with Virtual Channel Identifier (VCI) five, within a Virtual Path (VP). If the VC only carries signaling for its own VP then this configuration is referred to as associated signaling. If the VC carries signaling for other VPs then this is referred to as non-associated signaling, because the VPs are not directly associated with their signaling VC (see Figure 8.8).

The term "associated" can also be used to describe the signaling topology. In the associated signaling topology, the signaling takes a direct path between the two signaling points (see Figure 8.9). In the quasi-associated topology the signaling does not take a direct path between the two signaling points, but goes via an intermediate signaling transfer point, which is a packet switch that routes the signaling messages. Quasi-associated signaling is more complex than associated signaling, but it is more flexible and resilient.

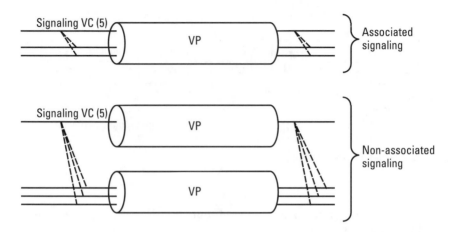

Figure 8.8 Associated and non-associated signaling.

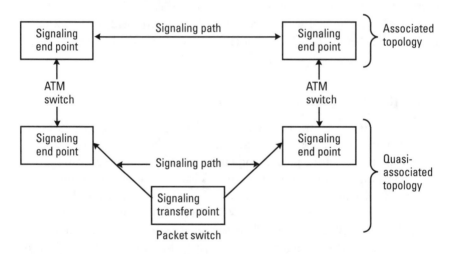

Figure 8.9 Associated and quasi-associated topologies.

B-ISUP is defined in four ITU-T recommendations, Q.2761 to Q.2764 [1]. Q.2761 is a functional overview. Q.2762 identifies the messages and parameters while Q.2763 specifies their format, using Abstract Syntax Notation One (ASN.1) [6]. Q.2764 defines the procedures used for calls and connections.

B-ISUP differs from N-ISUP in having modular application entities that are defined for each of its services. It has four services and they handle call control, bearer connection control, maintenance control, and unrecognized information. A new instance of an application entity is created for each new call or connection. These instances are identified by SIDs (Signaling Identifiers).

Although B-ISUP has the potential to separate call control and bearer control, these are combined in the main B-ISUP messages (see Figure 8.10). B-ISUP was also designed to have the potential to include new features through the modification of existing application entities and the definition of new ones.

B-ISUP messages link the UNI messages at either end of a signaling path. Information from the SETUP message from a calling party is carried in an IAM (Initial Address Message) that is sent from the originating ATM switch to the terminating ATM switch (see Figure 8.11). If the request in the IAM is accepted then an IAA (IAM Acknowledge) message is returned and a SETUP message is sent to the called party. An IAR (IAM Reject) message is returned if the request is rejected.

If the called party returns an ALERTING message then the terminating switch knows that all of the address information has been sent and returns an ACM (Address Complete Message) to the originating switch, which returns

Message	DIR*	Function
IAM: Initial Address Message	⟶	First message (destination number and connection characteristics)
IAA: Initial Address Acknowledge	⟵	Positive response to IAM Indicates resources reserved
IAR: Initial Address Reject	⟵	Negative response to IAM
ACM: Address Complete Message	⟵	Indicates receipt of all information needed to reach the called party.
ANM: ANswer Message	⟵	Indicates acceptance by the called party
REL: RELease	⟶⟵	Indicates a party wishes to release the call or connection.
RLC: ReLease Complete	⟵⟶	Indicates acceptance of a release request.

*DIR = Direction (initial switch - left: responding switch - right)

Figure 8.10 Main B-ISUP Messages.

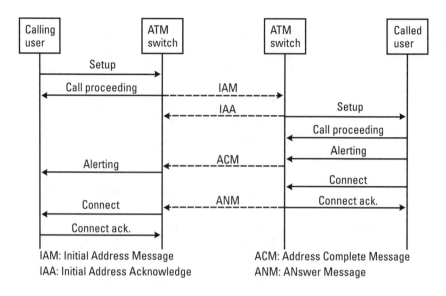

IAM: Initial Address Message   ACM: Address Complete Message
IAA: Initial Address Acknowledge   ANM: ANswer Message

Figure 8.11   Simple UNI and SS7 (B-ISUP) Message Flows: Call Establishment.

ALERTING to the calling party. Once the called party returns a CONNECT message the terminating switch knows that the connection has been accepted and returns ANM (Answer Message) to the originating switch. The originating switch then forwards CONNECT to the calling party.

Either party can initiate the release of a connection by sending RELEASE. The ATM switch that receives this sends an REL (Release) message to the other end of the network (see Figure 8.12). The remote switch responds by returning an RLC (Release Complete) message and by sending RELEASE COMPLETE to the remote party.

Routing for B-ISUP uses the MTP-3b (Message Transfer Part Level 3b) procedures for Signaling System No. 7 (SS7). The switches route using a hop-by-hop approach that requires them to be configured so they can determine the next switch on the route to the destination. There is no need for switches to have knowledge of the network topology or to distribute information about the network topology among themselves. This is in contrast to the source-based routing used with the ATMF's PNNI.

B-ISUP is not suitable as a universal inter-NNI because it does not use the AESAs defined for ATMF signaling although it can transport them transparently.

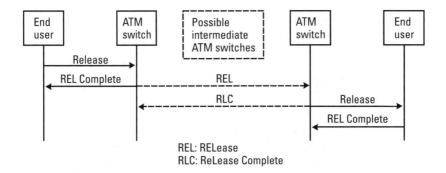

REL: RELease
RLC: ReLease Complete

Figure 8.12 Simple UNI and SS7(B-ISUP) Message Flows: Call Release.

8.4.2 ATM Forum Intra-NNI Signaling

The ATMF's intra-NNI, the PNNI, has a call signaling protocol that can specify routing and a separate call routing protocol that allows switches to determine the routes within the network.

8.4.2.1 PNNI Call Signaling

The PNNI call signaling protocol is a symmetrical version of the ATMF's UNI signaling. The SETUP message used in call establishment carries an information element that specifies the routing desired by the source of the signaling. Alternative routing can be specified in case a connection attempt fails.

In ITU-T's hop-by-hop routing, the network management is responsible for the configuration of routing information in each switch. This allows alternative signaling paths to be used if a signaling link fails after call establishment. In PNNI this is not possible because the route is specified at call set-up and the only flexibility is if a problem is discovered at that point and the call is 'cranked back'.

PNNI only uses the associated signaling topology, where the signaling messages are sent over a link that directly connects the signaling nodes. ITU-T's B-ISUP can use both associated and quasi-associated signaling topologies.

PNNI also uses the ATMF approach to QoS that associates parameters directly with connection types, whereas B-ISUP has the intermediate step of QoS classes. This is described in more detail in Chapter 19.

8.4.2.2   PNNI Routing

PNNI has a dynamic routing protocol that distributes topological information about the network among its switches, which collaborate to compute the routes. This is similar to the OSPF (Open Shortest Path First) approach for IP networks.

A network that uses PNNI is organized into a hierarchical structure of peer groups (see Figure 8.13). The peer groups at the lowest level of the hierarchy consist of ATM switches. At the higher levels the peer groups consist of logical groupings of lower peer groups. At the top of the hierarchy there is a single peer group.

Nodes in neighboring peer groups exchange routing information with their neighbors. These nodes discover the identity of their neighbors and the peer groups to which they belong by exchanging 'Hello' packets on VC 18 of each VP of each physical link. These packets also allow the nodes to determine the status of the physical links. Nodes that have links to nodes in other peer groups are called border nodes.

Nodes send PNNI Topology State Packets (PTSPs) to each other to indicate reachable destinations. PTSPs also indicate the bandwidth and QoS of the links between nodes. Nodes maintain routing databases, which they synchronize by sending and requesting database summaries.

Peer groups elect a peer group leader that has responsibility for representing the peer group at the next higher layer of the hierarchy. These leaders

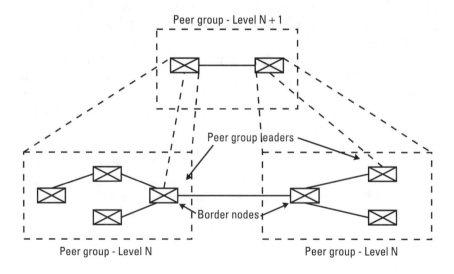

Figure 8.13   PNNI routing hierarchy.

exchange routing information over Switched Virtual Channel Connections (S-VCCs) that they establish between their border nodes.

Routes are either pre-calculated or calculated on demand. Pre-calculated routes are calculated by identifying possible destinations as a matter of course. This eliminates the need to calculate a route rapidly when it is required for a call. On-demand routes are calculated when there is no known pre-calculated route, and calculating them rapidly can be difficult. If a node cannot accept a connection then the call may need to be 'cranked back' and tried again.

8.4.3 ATM Forum Inter-NNI Signaling

The ATMF based its development of its inter-NNI, the B-ICI, on the B-ISUP part of ITU-T's SS7, but there are significant differences in the names and formats of the messages. Like the other ATMF standards, B-ICI specifies the full interface and not just the signaling. Ironically, the greatest threat to the adoption of the ATMF's B-ICI is ATMF's PNNI, because this is also used in practice as an inter-NNI.

8.5 Summary

ATM connections are more complex than narrowband connections because of their range of bandwidths, transfer capabilities, quality of service parameters and connection topologies. The ITU-T tackled this complexity by defining Capability Sets, but this was not effective and a pragmatic approach was subsequently adopted, which resulted in the Q.2931 standard for the User-Network Interface (UNI) and the Q.2761 to Q.2764 standards for the Network-Network Interface (NNI).

The ATM Forum (ATMF) has always taken a more pragmatic approach, and the result is its own UNI specifications, especially versions 3.1 and 4.0, that are similar to Q.2931. The ATMF has also produced NNI specifications, both for use within a private network (PNNI) and for carriers between networks (B-ICI). The differences between the ITU-T and ATMF specifications make interworking difficult.

Both bodies support bearer services, teleservices and supplementary services. Bearer services correspond to the connection types but may include ATM adaptation. Teleservices, such as teleconferencing, include signaling. Supplementary services are subscription options such as calling and connected line identification services. However, the ATMF signaling can use a larger address range, and is better for multipoint connections.

The different versions of the UNI signaling are similar and are based on narrowband ISDN signaling. Although the ATMF versions do not support some of the ITU-T messages, they do support additional features. In particular the ATMF specifications support proxy signaling, which allows connections to be established on behalf of another party.

The NNI standards are very different. The ATMF based its PNNI standard on its UNI signaling and added a protocol that communicates routing information throughout the network. ITU-T based its approach is based on its existing Signaling System No. 7 (SS7) with hop-by-hop routing. Although the ATM Forum has developed the B-ICI standard based on SS7, there are significant differences and its own PNNI standard is often used as an alternative.

Network operators are reluctant to support both the ITU-T and the ATMF specifications. They often show a preference for ATMF specifications because these are simpler yet more extensive, but there has also been a reluctance to introduce ATM signaling, especially since its relationship with the Internet has been unclear.

References

[1]   *B-ISDN Digital Subscriber Signaling System No 2, User to Network Interface Layer 3 Specification for Basic Call / Connection Control,* ITU-T Recommendation Q.2931, 1995.

[2]   *B-ISDN Signaling System No 7, B-ISDN User Part (B-ISUP),* ITU-T Recommendations Q.2761 to Q.2764, 1995.

[3]   *ATM User Network Interface Specification (Versions 3.1, 4.0 and Addendum),* ATM Forum, af-uni-0010.002, 1994 / af-sig-0061.000, July 1996 / af-sig-0076.000, January 1997.

[4]   *Private Network-Network Interface Specification (Version 1.0 and Addendum),* ATM Forum af-pnni-0055.000, March 1996 / af-pnni-0075.000, January 1997.

[5]   *Broadband Inter-Carrier Interface (Version 2.1),* ATM Forum af-bici-0068.000, November 1996.

[6]   *Specification of Abstract Syntax Notification One (ASN.1),* ITU-T Recommendation X.208 (= ISO 8824), 1988.

9

Management of ATM Switches

For Tweedledum said Tweedledee had spoiled his nice new rattle.
—Lewis Carroll, *Through the Looking-Glass*

One of the triumphs of cooperation in ATM is the agreement between the ATM Forum (ATMF) and ITU-T on the CMIP management model for ATM switches [1], despite their differences over ATM signaling and architecture. Perhaps it is sensible to celebrate such triumphs quickly because it is never clear how long they will last. In this case the persistent differences about ATM connections, and the reluctance to implement CMIP management models makes further celebration less likely.

Another reason for caution is that, although there is universal agreement on the basic CMIP model, there are differences between Europe and the US on the modeling of supplementary services. Some might say that the Europeans had spoiled the nice new model.

9.1 Background

Modeling of ATM switches began in parallel in ETSI (where it was soon incorporated into work in ITU-T) and in the ATMF. The intention was to extend the earlier CMIP model for ATM cross-connects. Fortunately the emphasis in the two bodies was different and complementary, and there was a willingness to co-operate.

This led to global acceptance of the ITU-T model, which covers ATM interfaces, ATM services and service adaptation, and the configuration of ATM routing.

9.2 ATM Interfaces

Signaling for ATM connections on demand is carried both on User-Network Interfaces (UNIs) and on Network-Node Interfaces (NNIs).

9.2.1 Modeling of UNIs

An ATM UNI is more complex than it first seems because of the transmission path to the ATM switch. The ATMF took the pragmatic view that the transmission path was simple and could be ignored. ITU-T decided to take a more general view that allows for VP cross-connection between the user and the switch that handles the signaling.

There can also be several signaling channels between the user and the ATM switch, and these can carry different signaling protocols. The signaling can be associated (when a signaling VC controls the connections within its own VP) or non-associated (when a signaling VC controls the connection in other VPs).

The concept of a UNI access was developed to model this complexity. A UNI access is a group of VPs from an end user that has the same protocol and the same type of signaling (associated or non-associated). If the UNI access has non-associated signaling then it has only one signaling channel, but if it has associated signaling then there is a signaling channel within each VP.

If the UNI access has associated signaling then it points to pairs of objects. Each pair consists of the end-point of a VPC and the end-point of the VCC that carries its signaling. The end-point of the VPC is modeled by an sVpTtp object [1] and the end-point of the signaling channel is modeled by a vcTTP object (see Chapter 5 and Figure 9.1).

The UNI access also identifies the profile of the signaling adaptation layer and associates a Virtual Path Connection Identifier (VPCI) with each of its VPCs. The signaling uses the VPCI to refer to the VPC regardless of physical interfaces. If the user can screen callers or connected lines then the UNI access contains an appropriate data object. The modeling of SAALs,

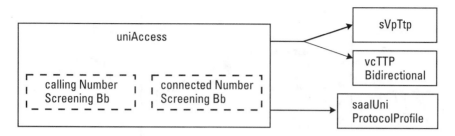

Figure 9.1 UNI access with associated signaling.

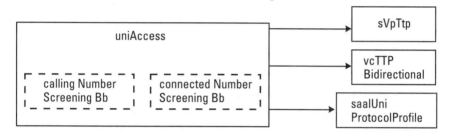

Figure 9.2 UNI access with non-associated Signaling.

VPCIs, and screening is the same regardless of whether the UNI access has associated or non-associated signaling.

If the UNI access has non-associated signaling then it points to all of the VPCs controlled by the signaling channel (see Figure 9.2), and identifies this signaling channel with another pointer.

9.2.2 Modeling of NNIs

An NNI access can be even more complex that a UNI access because ATM switches can interwork with narrowband switches. The link between an NNI and its VPCs is also less direct because the VPCs are gathered into groups with similar characteristics to simplify routing (see Figure 9.3).

Narrowband connections are also grouped at an NNI. In this case there are two levels of grouping because the lowest level, that for 64 Kbps connections, correspond to the VC layer in the ATM architecture. The narrowband end-points are modeled with crCircuitEndPointBb (call routing circuit end-point for broadband) objects, which identify the end-points of 64 Kbps circuits. A group of these end-points is at the same level in the architecture as a VP.

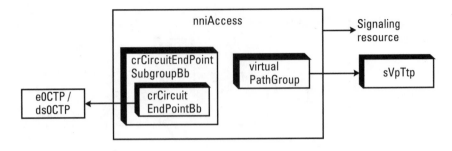

Figure 9.3   NNI access.

9.3   Service Profiles

The customer service profile links the customer's directory addresses to the customer's services. There is variation in the modeling of customer profiles, partly because of differences between customers, and partly because of differences between Europe and the US that the rest of the world stood back from. Even so, there are certain common features.

The customer profile always identifies the customer's access or accesses (see Figure 9.4) and the customer's directory address or addresses. If the customer has several addresses then the profile must contain the multiple number supplementary service (see Figure 9.5). The profile must also contain the data that indicates a special user, a payphone, or the like.

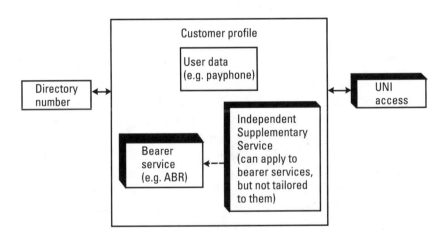

Figure 9.4   Multiple accesses, single directory address.

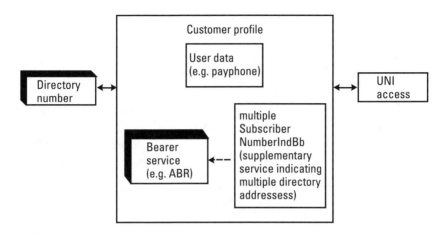

Figure 9.5 Single access, multiple directory addresses, services not customized by address.

The profile also contains the customer's other services. These are bearer services, telecommunication services (teleservices) and supplementary services (see Chapter 8). These services can be customized by directory address.

9.3.1 Bearer Services, Teleservices, and Supplementary Services

ATM bearer services correspond to the defined ATM connection types. The CMIP model [1] defines objects for the following services (see Chapter 19 for more details):

1. Available Bit Rate (ABR);
2. ATM Block Transfer with Delayed Transmission (ABT-DT);
3. ATM Block Transfer with Immediate Transmission (ABT-IT);
4. Deterministic / Constant Bit Rate (DBR / CBR);
5. Statistical / Variable Bit Rate (SBR / VBR);
6. Unspecified Bit Rate (UBR).

Although the model does not define objects for any specific teleservice, it does define objects for supplementary services. These can only be used in conjunction with a base bearer service or teleservice. Independent supplementary services are not specific to any particular base service. There is agreement that the following supplementary services are independent:

1. Closed User Group (CUG);
2. Direct Dialing In (DDI);
3. Multiple Subscriber Number (MSN).

Unfortunately there is a difference of opinion between Europe and the United States about the other supplementary services. The U.S. view is that these are dependent supplementary services, i.e., tailored to the customer's bearer service. The European view is that these are independent supplementary services, since they are defined independently of the base services, but they can be applied to different bearer services. The U.S. view is reflected in the ITU-T standard [1], but this was designed so that it could be restructured to form the European Standard [2]. The supplementary services that are affected are:

1. Calling Line Identification Presentation (CLIP);
2. Calling Line Identification Restriction (CLIR);
3. Connected Line Identification Presentation (COLP);
4. Connected Line Identification Restriction (COLR);
5. Closed User Group Subscription Option;
6. Subaddressing;
7. User-to-user Signaling.

9.3.2 Modeling of Supplementary Services

Other independent supplementary services are contained in the profile in the same way as the Multiple Subscriber Number (MSN) service (see Figure 9.5). It is possible for an independent supplementary service to point to base services, but only to indicate that it applies to them, and it should not be tailored to them.

Dependent supplementary services are not directly contained in the profile, but in the base service for which they are tailored. This approach is used in the US modeling of the CLIP, CLIR, COLP, and COLR services. These service allow calling or connected lines to be identified or their identification to be withheld (see Figure 9.6).

In the European model, the identification services are directly contained in the customer profile (see Figure 9.7). The independent identification services then point to the base services that they apply to.

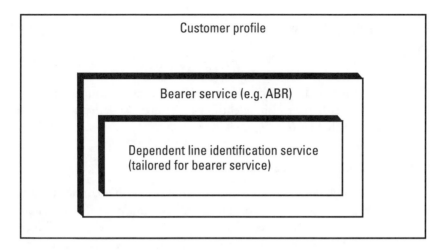

Figure 9.6 U.S. modeling of line identification services.

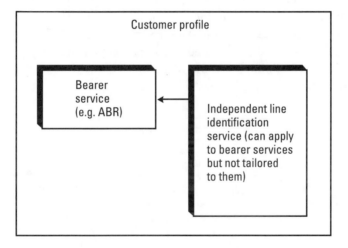

Figure 9.7 European modeling of line identification services.

The difference between the United States and the European modeling is as much driven by considerations of implementation as of the tailoring of supplementary services. In the U.S. model there is a supplementary service object within each base service that needs it. In the European model there is a single common supplementary service for several base services.

9.3.3 Customizing Services for Addresses

All of the services and user data in the customer profile apply by default to all of the customer's addresses. To constrain services or user data to specific addresses a customized resource object is added. Each customized resource object (customizedResourceBb) identifies an address and the services and user data that apply to it (see Figure 9.8).

9.3.4 The Circuit Emulation Service

ATM circuit emulation allows a VCC to act as a link in a connection that is set up for another service. Additional configuration is needed for circuit emulation because it includes configuration of the AAL at either end of the connection. AAL1 is normally used because the client services often require a constant rate connection with a timing reference.

Additional relationships are needed at the end-points of a VCC that is used for circuit emulation (see Figure 9.9). A pointer is needed to an AAL profile (see Chapter 7) to indicate the adaptation parameters. A second pointer is needed to a Circuit Emulation Service (CES) profile that indicates the size of the buffer that is needed to compensate for the ATM Cell Delay Variation (CDV). Pointers are also needed to identify the end-points of the client links that are served by the ATM connection.

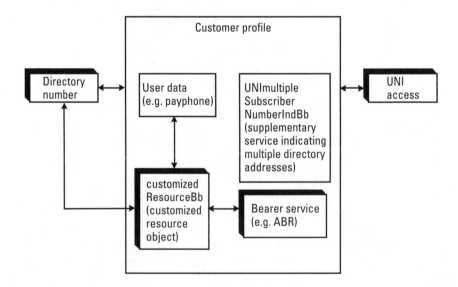

Figure 9.8 Bearer services customized by directory address.

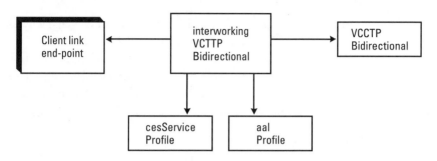

Figure 9.9 Configuration of circuit emulation end-points.

9.4 Configuration of the Routing Algorithm

Unlike the ATMF approach [3], the ITU-T approach to signaling between ATM switches [4] does not distribute routing information. In the ITU-T approach, the network operator must configure the routing algorithm in each switch.

Each switch has its own international code that identifies it when it initiates a connection. This code is modeled as by a callRoutingOfficeData object (see Figure 9.10).

An ITU-T E.164 address (see Chapter 8) has a prefix code that indicates whether the destination is local, national, or international. Each destination type is modeled by a prefixDigitAnalysis object, which has a fixed attribute for its type and a modifiable attribute for the corresponding set of prefix codes.

The routing analysis can take account of the origin of the call, the nature of the called party, and the carrier. The outcome of this analysis is

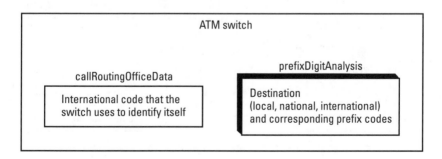

Figure 9.10 ATM switch identification and prefix tables for destinations.

identified by an analysisCriteria object, which holds a set of call characteristics and points to any relevant carrier data (see Figure 9.11).

Carriers can be specified by the call signaling. Each carrier is identified by a unique code that is either part of the ITU-T carrier plan or part of a national plan. Sometimes a carrier requires the Carrier Identification Parameter (CIP) and the Carrier Selection Parameter (CSP) to be signaled onwards to the next switch.

9.4.1 Modification of Destination Addresses

The final outcome of the routing analysis is a local or an abstract (i.e., remote) destination. In some cases there is an intermediate outcome that the address of the destination needs to be changed. This may be because an abbreviation was used or because of the nature of the outgoing route. The address can be changed by suppressing existing digits, by inserting new digits, or by replacing old digits with new.

These changes are modeled by a digit manipulation object. This object has attributes that indicate how digits are changed and what the new destination type is.

9.4.2 Local Destinations

If the analysis indicates a local destination then the call is not routed to another switch. Valid local destinations are defined by first specifying an initial set of numbers and then specifying a subset of excluded numbers. The initial set and the excluded subset are both defined by listing their most

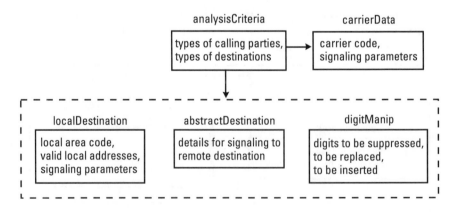

Figure 9.11 ATM routing analysis.

significant digits. The time for which the local destinations will be rung can also be defined.

The local area code, the valid local addresses, and the ring time limit are modeled by attributes of a local destination object.

9.4.3   Abstract (Remote) Destinations

Remote destinations are abstract because the destination switch is unknown as ITU-T specifies hop-by-hop routing. An abstract destination may have a limit on the size of the called address it can handle and the time for which the switch will attempt to place the call to it.

The routes to an abstract destination are determined by a post analysis evaluation.

9.4.4   Routes to Remote Destinations (Post Analysis Evaluation)

The requirements of different calls must be taken into account when abstract destinations are linked to routes. For each abstract destination, lists of routes must be specified that meet certain criteria (see Figure 9.12).

The range of bandwidth supported in each direction must be specified for the list of routes. The transit delay range and ATM bearer capabilities must also be specified. The routes to the abstract destination may also be able

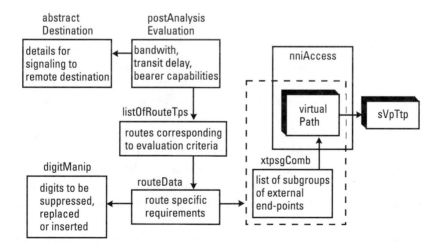

Figure 9.12 Evaluation and route identification.

to support narrowband capabilities, in particular the narrowband ISDN audio, digital and video capabilities.

A post analysis evaluation object links an abstract destination with a list of routes, and holds the corresponding set of criteria. Each route consists of a collection of termination points with similar properties. The route may require manipulation of the destination address, or specific information to be sent to the next switch. There may also be a minimum number of digits that need to be signaled before the route is seized.

A route data object has attributes that correspond to the characteristics of a route and a pointer to indicate the associated group of connections at the NNI. This pointer either identifies a single group of end-points or a list of ordered subgroups. A list of subgroups is modeled by an external termination point subgroup combination (xtpsgComb) object.

9.5 Summary

Despite differences over signaling and architecture, ITU-T and the ATMF have been able to agree on a CMIP management model for ATM switches. However, differences about the modeling have arisen because Europe and the US take different approaches to the modeling of supplementary services.

UNIs are modeled as UNI accesses. A UNI access is a bundle of VPCs that have the same type of signaling. If associated signaling is used then each VPC contains its own signaling VCC. If non-associated signaling is used then the UNI access has a single signaling VCC that controls all of the connections for its bundle of VPCs.

A customer profile links UNI accesses, directory addresses and communication services. The profile contains specific user data and the services specific to the customer. Customer profiles also allow services to be customized for different addresses.

These services are the bearer services, which correspond to connection types, and supplementary services such as caller identification. Teleservices, which involve signaling, remain undefined in the model.

Supplementary services are dependent if they are tailored to a base service. Dependent supplementary services are contained within their base services. Independent supplementary services are contained directly within the customer profile and identify the base services to which they apply. Europe and the United States take different views about which supplementary services are dependent and which are independent.

NNIs are modeled as NNI accesses. Since connections are routed over NNIs, an NNI access is represented as a collection of groups of VPs that have similar characteristics. As NNIs on ATM switches may also interwork with the narrowband network, an NNI access can also consist of a number of narrowband connections. In this case there is yet a further layer of bundling because basic 64 Kbps narrowband connections need to be bundled together in the same way that VCCs are bundled together in a VPC.

The configuration of the routing algorithm in an ATM switch is not necessary for ATMF signaling between switches because this signaling distributes routing information. For ITU-T hop-by-hop routing, an algorithm is configured to identify whether the destination is local or remote, and whether or not the called number needs to be manipulated. Following this analysis the connection requirements must be evaluated so that a suitable VPC can be found.

References

[1]   *Broadband Switch Management,* ITU-T Recommendation Q.824.6, 1998.

[2]   *ATM Switching Network Element: Q3 Interface Specification,* ETSI Standard EN 301 064-1, 1998.

[3]   *Private Network-Network Interface Specification (Version 1.0 and Addendum),* ATM Forum af-pnni-0055.000, March 1996 / af-pnni-0075.000, January 1997.

[4]   *B-ISDN Signaling System No 7, B-ISDN User Part (B-ISUP),* ITU-T Recommendations Q.2761 to Q.2764, 1995.

10

Internet Communication

Professionals built the Titanic. Amateurs built the Ark.
—Anonymous

There is a lesson in the success of the Internet, for those who have eyes to see and ears to hear. For those who have minds to think, there may be an even deeper lesson. The Internet was not developed formally by established communications authorities. On the contrary, the Internet Engineering Task Force (IETF) is proud of its anarchistic heritage.

But yesterday's anarchists can easily become today's bourgeoisie. There are signs of conservatism in the slow rate of acceptance of revisions to established Internet practices. Amateurs may have built the Ark, but it was their descendants who built the Titanic.

10.1 Introduction

The term "Internet Protocol" (IP) is often used to refer to the constantly evolving suite of Internet protocols [1], rather than just the specific standard that defines the structure of the IP datagrams that are the fundamental units of IP communication. The suite of Internet protocols supports both assured and non-assured communications and these in turn support a wide and growing range of applications protocols (see Figure 10.1). Most IP communication uses IP version 4 (IPv4) [2], which is the basis of this chapter and the immediately subsequent chapters. IP version 6 (IPv6) [3] is discussed later in the final part of this book.

IP communication is connectionless. This means that IP datagrams contain their own destination addresses and they are individually routed to these destinations. To be accurate, it should be noted that IPv6 includes a mechanism to establish paths for datagrams and it is possible to create paths for IPv4, but this is exceptional.

IP communication is decoupled from the physical hardware that supports it. In this respect IP is like ATM. In a physical network such as an Ethernet, frames carry the individual IP datagrams. Frames also carry the address registration messages that co-ordinate the mapping between the frame addresses of the physical network and the logical IP addresses. The frames label the information that they carry so that IP datagrams and address registration messages can be differentiated.

IP datagrams have a header and a payload (see Figure 10.2). The protocol field in the header indicates the nature of the payload. Some of the most common values are those for messages that belong to the ICMP (Internet Control Message Protocol), the Transmission Control Protocol (TCP), and the User Datagram Protocol (UDP).

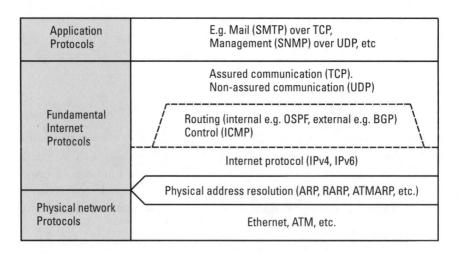

Figure 10.1 The Internet Protocol architecture.

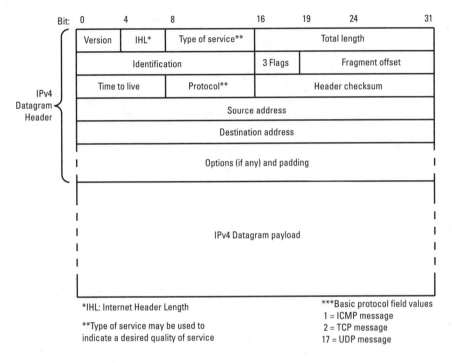

*IHL: Internet Header Length

**Type of service may be used to
indicate a desired quality of service

***Basic protocol field values
1 = ICMP message
2 = TCP message
17 = UDP message

Figure 10.2 IPv4 datagram format.

10.2 IP Addresses and Address Resolution

IP datagrams contain IP addresses that are used for communications across different networks. IP addresses have to be mapped onto the link-level addresses that are used over the supporting communications networks, suchas Ethernet, frame relay and ATM, that carry the IP datagrams. These link-level addresses may be included in the frames transmitted over a packet network or they may identify connections over a connection-oriented network.

It is possible to tell if two machines are on the same network by examining the network part of their IP addresses (see Figure 10.3). An IP address has a variable length address type prefix. This prefix defines the location and size of the network and local parts of the IP address. The prefix can also indicate special addresses, such as an address for multicasting to multiple destinations.

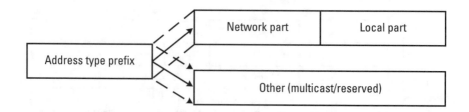

Figure 10.3   Structure of IP addresses.

The local part of an IP address can be split into a subnet address and a machine address. A collection of subnets that have the same network partintheir IP addresses appear as a single network to the rest of the Internet. Routing between subnets is simpler than routing across the entire Internet because the subnet address specifies a particular part of the local network.

A machine on a network that is partitioned into subnets needs more than the IP address of a destination to decide whether or not a datagram should be sent to a router [4]. This is because the datagram still needs to be sent to a router even when the destination machine is on the same network, if the destination machine is on a different subnet.

The sending machine needs information about which part of the IP address corresponds to the address of the subnet, because the machine cannot determine this from the IP address alone. The sending machine needs the mask that identifies the part of the IP address corresponding to the subnet address.

An address server is often used to manage the IP addresses on a network. It is possible to assign IP addresses to machine manually, but this is not especially efficient. An address server is useful for translating between the link-level addresses used on the network and the IP addresses used in datagrams. An address server is essential if the network does not support broadcasting since it provides a commo nreferenc node when there is no simple way to send a message to all machines on the network.

10.2.1   The Address Resolution Protocol (ARP)

The Addresses Resolution Protocol (ARP) is a simple solution to the problem of determining the link-level address of a destination machine from its IP address when a network allows messages to be broadcast. The ARP must accommodate the different types of link-level addresses that are used on different networks [5]. The ARP approach to address resolution is to broadcast a request containing the IP address to all of the machines on the network (see Figure 10.4). The machine that knows the correct mapping responds with the requested link-level address. If the network does not allow a message to be broadcast to all of its machines, then a different approach is needed, such as the one described later for ATM.

A broadcast ARP request also contains both the sender's IP address and the sender's link-level address. The logical structure of ARP messages and Reverse ARP (RARP) messages is given in Figure 10.5. RARP requests allow a machine to determine its own IP address if it does not know it, or is unable

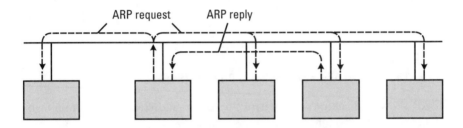

Figure 10.4 Determining the link-level address for an IP address.

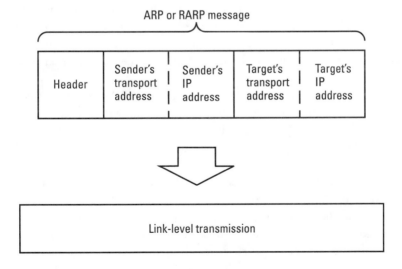

Figure 10.5 Logical structure of ARP and RARP messages.

to remember it. A field in the message header differentiates between ARP and RARP messages.

All of the machines that receive an ARP request can add the mapping between the IP and link-level addresses of the sender to their own address table. They are able to do this because the information is contained in the request. They ought to do this to avoid the network overhead of an unnecessary broadcast ARP request later.

A machine that receives an ARP request containing its own IP address adds its link-level address to the message and returns it to the sender. This link-level address is added to the reply message to ensure that the mapping is always available to the IP software.

It is possible for a router to respond to ARP request for remote destinations by returning its own link-level address. This approach is known as proxy ARP because the router acts as the proxy for ARP requests for remote destinations.

10.2.2 The Reverse Address Resolution Protocol (RARP)

RARP allows a machine on a network to ask an address server what its own IP address is. It also allows a machine to discover the IP addresses of other machines on the same network.

An ARP request contains the IP address of a machine and an ARP reply also contains the link-level address of that machine. A RARP request contains the link-level address of a machine and a RARP reply also contains the IP address. The reply to a RARP request can return the IP address of a third machine because the address server sends the IP address corresponding to the link-level address specified in the request back to the machine that sent the request.

A RARP request can be broadcast like an ARP request because the link-level address of an address server may not be known. A reply to a RARP request, unlike the reply to an ARP request, may be sent by more than one machine. This is because there may be several machines acting as address servers.

The possible multiple replies to a RARP request need to be coordinated to avoid congestion. A simple way to do this is if each address server adds arandom delay before sending its reply. This approach does not require coordinated configuration of the different servers. A more sophisticated approach is to configure servers as primary or secondary servers. A primary server replies to an RARP request immediately. A secondary server does not reply to a RARP request unless it is repeated.

10.2.3 Multicasting on the Internet

Although at one extreme multicasting is equivalent to one-to-one communication, at the other extreme it is not equivalent to broadcasting because participants in multicasting belong to a well-defined group. In order to multicast on the Internet, it is necessary to define the members of the multicasting groups, and to identify the routes between them. The Internet Group Management Protocol (IGMP) [6] handles the membership of multicasting groups. The coordination of routers for multicasting is less well defined than that for normal IP communications. The coordination of routers for normal IP communications in turn is not governed by a single common approach.

IP multicasting is simplified when it is confined to one network since the problems of routing do not then apply. It is further simplified if the network protocol also supports multicasting. If the network protocol does not adequately support multicasting then multicast IP datagrams must be sent separately to all of the appropriate link-level addresses, and how this is achieved is specific to the particular type of network.

IP multicasting is possible because IP has a class of destination addresses reserved for multicast groups. Each of these addresses is not associated with a single machine, but with a group of machines. All of the machines belonging to a particular multicast group receive the datagrams that are sent to the group's destination address. Other machines do not need to belong to the group to send to the group's members.

The Internet has both permanent groups and transient groups. Permanent groups have addresses that are well known and exist regardless of the machines that join or leave them. Transient groups have dynamically assigned addresses and cease to exist if all their members leave.

To join a group, a machine declares its membership in an IGMP message that is sent to the special all-hosts multicast address. This is the second lowest multicast address and the lowest usable address. IGMP messages that are addressed to the all-hosts group are not forwarded beyond the network where they originated.

All of the machines on a particular network that use IGMP are members of the all-hosts group for that network. Gateway routers can use membership of the all-hosts group to advertise themselves. Host machines can use can use membership of the all-hosts group for address resolution.

The initial membership declaration is sent to the all-hosts address on the local network because this ensures that multicast routers are aware of the membership without needing to belong to the group themselves. This avoids the need for every multicast router to belong to every multicast group and having the burden of receiving every multicast datagram. Sending initial declarations to the all-hosts group also allows multicasting to be supported when there are no multicast routers, but this is not recommended because multicast routers are needed to confirm that machines have not subsequently left the group.

When a multicast router on a network receives a declaration of membership in a new group it notifies other multicast routers throughout the Internet that it needs to receive multicast IP datagrams for the new group. A multicast router also polls to check that every group still has at least one member on its network. To reduce the number of polling messages, routers send a single message to ask about all group memberships, and members do not reply if they hear a reply from another member. IGMP queries are addressed to the all-hosts group because they are not specific to a particular group.

Host machines reply to these polls with IGMP reports that are staggered over time. Each report is similar to the initial declaration of membership, but it is addressed to the particular group and not to the all-hosts group. Like the initial declaration it is only send on the network that originated it, but this is achieved by setting the time-to-live field to one. This means that only group members on the network can hear it and they then know that the query has been answered for their group and so do not also reply. If no host answers then the router knows that there are no local members and can inform other routers to cease forwarding datagrams for the group. Membership of the all-hosts group is not reported.

In IGMP version one, members leave the multicast groups passively by ignoring IP datagrams that are sent to the group address and by ceasing to confirm that they are members when polled by routers. In IGMP version 2 there is an explicit message that allows hosts to declare that they have left the group.

10.2.4 Advanced Address Registration: BOOTP and DHCP

A RARP request needs to include the link-level address that corresponds to the unknown IP address, and this can be a problem if link-level addresses are assigned dynamically. It is also a problem for programmers working at the application level because it requires specific knowledge of the network protocols.

The alternative is to use an application level protocol. This is feasible because the request can be broadcast on the local network using IP. It canalso be more effective than RARP because it can include additional information to assist in bootstrapping when a machine needs to discover its own IP address.

The Bootstrap Protocol (BOOTP) is an IP application protocol that was developed as an alternative to RARP. The Dynamic Host Configuration Protocol (DCHP) was subsequently developed to allow for the dynamic allocation of IP addresses. Dynamic allocation of IP addresses is useful when portable computers need to be connected to different networks and when many computers may be connected to the same network but with only a small number connected at any time.

For simplicity, both BOOTP and DCHP use the User Datagram Protocol (UDP) for non-assured IP communications. The Transmission Control Protocol (TCP), which is more complex but provides assured communication, is not used. As a consequence, BOOTP and DCHP have to take account of the possibility of messages being lost.

10.3   IP Over ATM

Sending IP datagrams over an ATM connection is relatively easy because it just a matter of following the rules for ATM adaptation. Relating the proposed global ATM network to the Internet is less easy because they are alternative approaches to global communications.

The simpler problem is to make an ATM network suitable for use within the Internet. The ATM Forum proposed LAN Emulation (LANE) [7, 8] for this, using ATM applications that make a connection-oriented ATM network mimic a connectionless network. This approach has now been largely discounted because it requires servers for address resolution and broadcasting and does not capitalize on the strengths of ATM.

The alternative approach developed by the Internet Engineering Task Force (IETF) is to group a number of machines on an ATM network into a Logical IP Subnet (LIS) (see Figure 10.6). This approach has become known as classical IP over ATM [9]. Each machine in an LIS has the same network part in its IP address. The members of an LIS are allowed to send IP datagrams to each other over the ATM network. Classical IP over ATM does not support broadcasting.

A primary requirement of classical IP over ATM is to modify ARP for operation on ATM networks. ARP was designed for networks that support broadcast communication. ATM was not designed for broadcast communication. ATM was primarily designed for point-to-point communications.

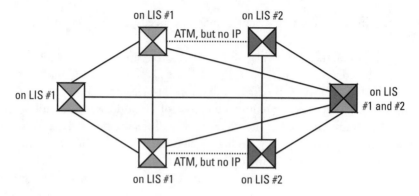

Figure 10.6 Logical IP subnets (LISs) on an ATM network.

Multipoint or broadcast communication was a secondary consideration for ATM and it is often not supported.

A machine that belongs to an LIS needs to know where on the ATM network to send a datagram with a particular destination IP address. The machine cannot find the ATM address of the destination by broadcasting an ARP request because most ATM networks do not support broadcasting. To find the ATM address of the destination each member of the LIS needs to know the location on the ATM network of an ARP server.

Classical IP over ATM requires host machines on different LISs to communicate through a router to keep the number of ATM connections required down to a manageable level. Unfortunately this means that if an ATM connection is possible between host machines on different LISs then it cannot be used for IP traffic.

10.3.1 ATM Link-Level Addresses

There are two types of ATM link-level address because there are two ways of making ATM connections. A Permanent Virtual Connection (PVC) is created in response to one or more configuration requests from the network management system. A Switched Virtual Connection (SVC) is established and released in response to user signaling in ATM calls.

Regardless of the type of ATM connection, the sending machine also needs to know what cell identifiers to put in the ATM cell headers. This is because the header of each ATM cell on a physical connection has a field that identifies the VP (Virtual Path) to which it belongs and a field that identifies its VC (Virtual Channel) within that VP.

If the only connections in an LIS are PVCs then cell identifiers may be all that is needed to send datagrams to a destination machine. This is because the sending machine can rely on the ATM network to route the cells that carry a datagram to the appropriate destination according to the labeling of the cells.

If the ATM connections are SVCs then the sending machine needs to set up a connection to the destination machine. The connection is established and released using messages that are send over a signaling channel to an ATM switch. To establish the connection, the sending machine needs to know the ATM address of the destination.

Unfortunately, one of the many things that the ITU-T and the ATM Forum have not been able to agree on is the format of the ATM addresses for SVCs. Although the address in the cell headers can be used as a link-level address for PVCs on the same physical interface, there is also no agreed format for link-level addresses for PVCs that can be used across a network.

An ATM SVC address can be the combination of an eight octet ITU-T E.164 number and a 20 octet ATM Forum Network Service Access Point (NSAP) address. The E.164 number is specified for public networks while the NSAP address can be used on private networks. It is tempting to think of the NSAP address as an extension, similar to an extension on a PABX, but this is a bit misleading because the private network could be larger than the public network and because both E.164 numbers and NSAP addresses can be used as global addresses.

10.3.2 The ATM Address Resolution Protocol (ATMARP)

The ATM Address Resolution Protocol (ATMARP) takes account of the two formats of ATM addresses by using a split address format. In ATMARP, the complete ATM address consists of a primary address plus a sub-address. This approach allows a machine to have a local NSAP address on a private network that has in turn an E.164 number on a public network. However, the address structure in ATMARP makes no assumption about whether the primary address is an E.164 number or an NSAP address. Each part of the complete address has an associated field that indicates its type and size. If there is no sub-address then the associated size field takes the value zero.

ATMARP messages also have a field that indicates their type. In addition to requests for the ATM address that corresponds to an IP address, there are also inverse ATMARP requests [9, 10].

An inverse ATMARP requests asks the other side of an ATM connection to return both its ATM address and its IP address. This is possible because the ATM connection defines the destination of the request even when the destination ATM address or IP address is unknown. Inverse ATMARP requests are not the same as the Reverse ARP (RARP) requests because an RARP request for ATM would request the IP address that corresponded to the ATM address. There are positive and negative replies for each type of ATMARP requests. Negative replies are included to confirm that the request has been received.

Conventional ARP requests are broadcast and ask all machines on a network to examine the included IP address and if it is theirs to respond with their link-level address. ATMARP requests ask a server to return the address that corresponds to the included IP address (see Figure 10.7). A RARP request asks one or more servers to return the IP addresses that correspond to the included link-level addresses.

An ATMARP server validates the entries in its address table by sending inverse ATMARP requests to the machines that are connected to it. A server does not need to poll machines that have recently sent ATMARP requests because an ATMARP request already contains the link-level and IP addresses that would be send in response to an inverse ATMARP request.

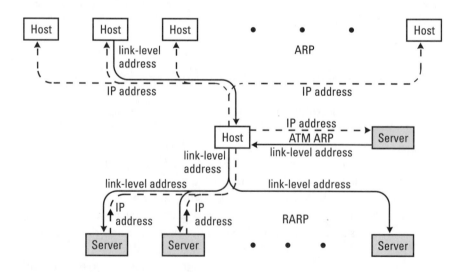

Figure 10.7 ARPs, ATMARPs, and RARPs.

10.4 Internet Control Messages

The Internet Control Message Protocol (ICMP) [11] is the mandatory protocol that is used between IP machines for testing, error reporting and control (see Figure 10.8). ICMP messages can be sent back to the original source of an IP datagram or on to other known destinations. The ICMP cannot send information back to intermediate routers because the route traveled by a datagram is not normally known.

Each ICMP message is carried as the payload of an IP datagram because datagrams can be sent across different physical networks. ICMP messages start with a four-octet header (see Figure 10.9). The type field indicates the particular type of ICMP message. The code field provides

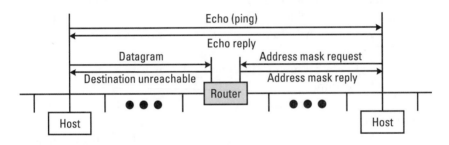

Figure 10.8 Example ICMP messages.

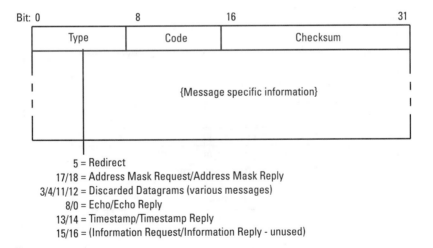

Figure 10.9 ICMP message format.

additional information that is specific to the different types of ICMP message. The checksum is calculated for the entire ICMP message using the same arithmetical approach that is used for the header of an IP datagram.

ICMP messages are used to control the local network, to report discarded datagrams, and for end-to-end control.

10.4.1 Control on the Local Network

Certain ICMP messages control the redirection of traffic on the local network and allow the local network to be partitioned into subnets.

10.4.1.1 Redirection Messages

The initial connection of a machine to a network is simplified if the machine only has to know the address of a single router. Even this is not essential if the router uses proxy ARP to masquerade as a remote destination. Although access to a single router gives a sending machine access to the Internet, it may not be ideal. This is because the local network may have a better router for a particular destination.

If the router that receives a datagram knows that the datagram should have been sent to a different router, it forwards the datagram so that the sending machine does not need to send it twice, but informs the sending machine of the other router in an ICMP redirect message (see Figure 10.10). The ICMP redirect message can have one of four values to specify different

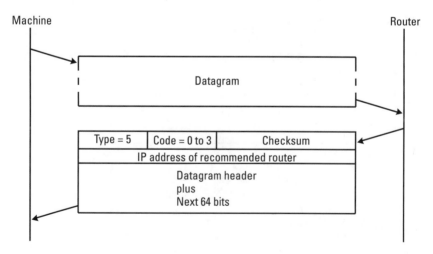

Figure 10.10 ICMP redirection.

situations. Redirection can be requested according to destination network or destination machine, and for all services or for a particular service. Destinations and services are identified in headers of datagrams.

Redirection of traffic using ICMP messages is restricted. Not only is it limited to the redirection of datagrams on the local network, the redirection message can only be sent from a local router. The ICMP approach also sidesteps the issue of how a router knows that a better router exists. These issues are addressed, often badly, by the Internet routing protocols.

10.4.1.2 Subnet Address Masks

To obtain a subnet address mask, a machine sends an ICMP address-mask-request to a router (see Figure 10.11). If the machine does not know the address of the router it must broadcast the request. The router returns an ICMP address-mask-reply message that contains the mask.

Both the request and the reply use the same message format and this includes the address mask field. The address mask indicates the bits within IP addresses that are used for subnet addressing for a particular network. The bits do not need to be contiguous, but this is usual because it is simpler.

10.4.2 Reporting Discarded Datagrams

An ICMP message is sent back to the source of a datagram if the datagram is known to have been lost. Datagrams can be lost without trace if, for example,

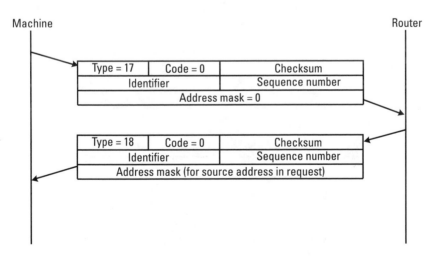

Figure 10.11 The ICMP Subnet Address Mask Transaction.

the destination machine is switched off and the destination network does not take account of this. The components of the Internet take a best effort approach to the delivery of datagrams, but they do not guarantee delivery or confirm delivery.

A machine that discards a datagram knows that the datagram is lost and sends an ICMP message back to the source (see Figure 10.12). The machine can send back one of four different types of ICMP message, depending on why it discarded the datagram. Each of the four message types has the same basic format, which includes diagnostic information. The diagnostic consists of the header and the first eight payload octets of the datagram that was discarded. The payload octets are included in the diagnostic because they contain key information about the protocol that sent the datagram.

A router may have discarded a datagram because the destination was unreachable. The reason why the destination was unreachable is indicated in the code field in the message header. Even if the destination is reachable, a datagram can still be discarded because of the expiry of a time limit.

A router may have discarded a datagram because its time to live, in hops, has expired. This can happen if a problem with routing tables has created a cyclic route from which datagrams cannot escape. The destination

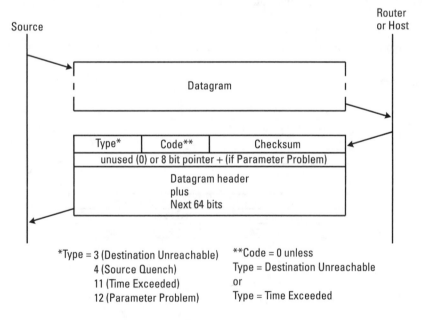

Figure 10.12 ICMP Discarded Datagram Report.

machine may also discard a datagram if it has started to reassemble it from received fragments and has not received all the fragments in the allowed time. Which time-out has expired is indicated in the code field of the time-exceeded ICMP message.

A router may also discard datagrams because it is congested. In this case, a quench ICMP message is returned to the sending machine. The message both notifies the sending machine that a datagram has been discarded and requests the sending machine to send less datagrams. The sending machine should reduce the rate at which datagram are sent, but it can increase the rate later so long as this does not provoke further quench messages.

A datagram may also be discarded if there is a problem with its content, particularly if it contains an invalid parameter. In this case a parameter-problem ICMP message is sent with a pointer that identifies the octet that caused the problem. If the problem is an optional field that is required but not included, then the code field in the message header is different (one, not zero) and there is no pointer.

10.4.3 End-to-End Control

ICMP messages are also used to check if a destination machine is reachable and for the estimation of time delays between machines.

10.4.3.1 Echo Testing (Pinging)

Checking that it is possible to reach a destination machine is one of the simplest and most effective ways of testing Internet communications. A host machine can send an echo-request ICMP message to a destination machine (see Figure 10.13), which returns an echo-reply message. An echo-request message carries an identifier and a sequence number to label the message. The echo-request may also carry data. The destination machine sends this information back to the original sender in the echo-reply message. This process is known as pinging.

10.4.3.2 Timestamping

It can be difficult to calculate the absolute value of the transmission delay for datagrams that are sent between two machines. This is because the clocks on the two machines are often not synchronized. Synchronization of the clocks is difficult in turn because the transmission delays vary with time and may be different in each direction. Synchronization of the clocks on all machines

Source
machine

Destination
machine

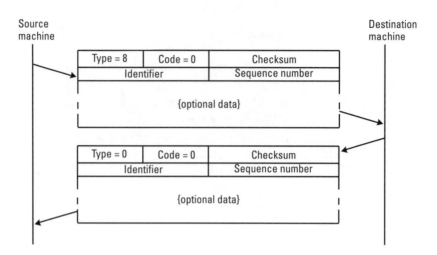

Figure 10.13 ICMP pinging.

would be even more difficult because a time reference would need to be agreed first.

It is easier to calculate the relative delays and the round trip delays between two machines because these do not rely on synchronization of the clocks on the machines. A machine that wishes to initiate the ICMP process for this sends a timestamp-request ICMP message to another machine, which responds with a timestamp-reply ICMP message (see Figure 10.14).

The format of the ICMP timestamp messages does not engender a particularly deep sense of awe. The timestamp-request contains identifier and sequence number fields that are returned in the timestamp-reply, but the timestamp-request also contains the time that it was sent and this too is returned, making the identifier and sequence number field redundant. The timestamp-request also contains fields for the time that it was received and for the time the reply was generated. These fields are also not appropriate in the request.

10.5   End-to-End Data Transport

IP datagrams are differentiated from ARP and RARP messages at the physical transport level. Often this is done by a field in the header of a physical frame that indicates the nature of the payload. There is a similar field, the

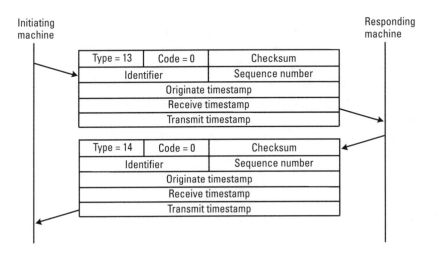

Initiating machine

Responding machine

Type = 13	Code = 0	Checksum
Identifier		Sequence number
Originate timestamp		
Receive timestamp		
Transmit timestamp		

Type = 14	Code = 0	Checksum
Identifier		Sequence number
Originate timestamp		
Receive timestamp		
Transmit timestamp		

Figure 10.14 The ICMP timestamp transaction.

protocol field, within the header of each IP datagram. This field indicates the Internet protocol that the datagram is carrying.

The Internet Control Message Protocol (ICMP) has already been mentioned. Two other protocols, the User Datagram Protocol (UDP) [12] and the Transmission Control Protocol (TCP) [13], are the key protocols that transport data for Internet applications protocols.

TCP provides assured communication. It transports data for the e-mail protocol, the Simple Mail Transfer Protocol (SMTP), and for other application protocols that prefer to use a relatively complex data transport protocol that provides assurance that the data has been delivered intact.

UPD is a simpler protocol than TCP. UPD carries data for the Simple Network Management Protocol (SNMP) and other application protocols such as the Bootstrap Protocol (BOOTP) and the Dynamic Host Configuration Protocol (DHCP) mentioned earlier. These application protocols prefer to use a simple data transport protocol and are willing to accept that some messages will be lost.

10.5.1   The User Datagram Protocol (UDP)

UDP datagrams are carried in the payload of IP datagrams. If the value of the protocol field in the header of an IP datagram is 17, then the IP datagram is carrying a UDP datagram.

UDP datagrams also have headers. A UPD header adds port addresses that extend the Internet source and destination addresses contained in the header of the IP datagram that carries it (see Figure 10.15). Port addressing allows different software applications to operate between the same pair of machines. A UDP header also indicates the length of the datagram and carries a simple arithmetical checksum.

The source port address indicates the port to which replies should be returned. If the value of the source port address field is zero then no source port is specified. Port addresses 161 and 162 are reserved for SNMP.

The possibility of not computing a checksum for the UDP header is catered for because calculating it involves one's complement arithmetic. The checksum is not a proper CRC that uses all possible values of the field. In one's complement arithmetic, the value zero can be represented in two ways (all bits zero and all bits one). All ones are used for zero when it is calculated, and all zeros indicate that the checksum has not been calculated.

The checksum in the UDP header violates strict layering because information in the IP header is used to form a pseudo-header for its calculation (see Figure 10.16). The Internet source and destination addresses in the IP header are used in the calculation of the UDP checksum as a simple way of confirming that the UDP payload has been carried between the correct Internet hosts. The UDP header identifies the ports of these hosts, but not the hosts themselves. The protocol field in the IP header is also used in the calculation as a simple way of confirming that the IP payload is indeed a UDP datagram.

UDP datagrams may be lost or they may be duplicated through retransmission. If they are delivered without duplication, then they may be delayed or not delivered in the sequence in which they were sent. The Transmission Control Protocol (TCP) was developed to solve these problems.

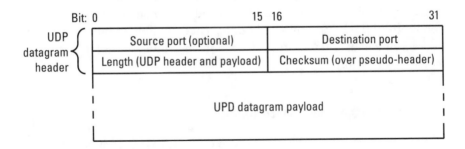

Figure 10.15 UDP datagram format.

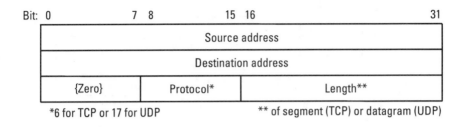

Figure 10.16 Pseudo-header for UDP and TCP checksum generation.

10.5.2 The Transmission Control Protocol (TCP)

The Transmission Control Protocol (TCP) controls the flow of information between machines and confirms that the information is transferred in the correct sequence and without loss or duplication. This means that TCP enables IP applications to be developed independently of the details of the communications. TCP differs from similar protocols that have been developed for telecommunications because it is a general-purpose protocol and is not specific to a particular form of call control signaling.

When TCP is carried over IP, the payload of each IP datagram contains a TCP segment. TCP segments, like UDP datagrams, have headers that identify ports at the source and destination machines. The headers also contain arithmetical checksums that are calculated in the same layer violating way as those in UDP (see Figure 10.16).

Unlike UDP datagrams, TCP segments contain sequence numbers (see Figure 10.17). These sequence numbers do not refer to the segments, but to the octets carried in the segments. Decoupling the octet stream from the segments that carry it is useful because of the flexibility it allows for retransmissions. Octets from a lost segment can be retransmitted as a number of smaller segments or merged with subsequent octets in a single large segment. Segments that are returned acknowledge the octets that have been received and contain flow control information. These returned segments can also carry a payload of application data.

Also, unlike UDP, TCP headers do not contain a length field. In UDP, the length contributes twice to the arithmetical checksum. In TCP, the checksum simply confirms length of the payload as given in the header of the IP datagram.

The segment headers contain flags that control the establishment of logical TCP connections between machines. These flags also control the normal release and the abnormal abortion of the connections.

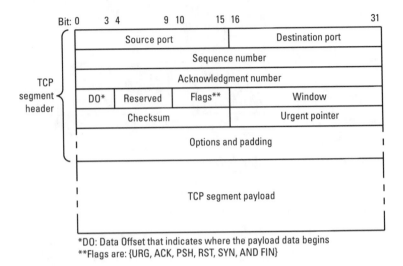

Figure 10.17 TCP segment format.

10.5.2.1 Data Acknowledgement and Flow Control

The acknowledgement flag in the header indicates whether or not the header's acknowledgement and window fields are valid. A valid acknowledgement field indicates the sequence number of the first octet that has not yet been received. A valid window field indicates the number of additional octets that the receiver is prepared to accept. If the value of the window field is zero then the window is closed and the suspension of transmission is requested. Segments that contain urgent data or that check for the loss of the message to reopen the window can still be sent even if the window is closed.

This approach is more effective than acknowledging each segment that is received because fewer acknowledgements are needed. It is also simpler than explicitly identifying the gaps in the received information, but it can be less efficient that identifying the gaps because it requires all subsequent data to be retransmitted once data is lost.

A header length field is used to identify the start of the segment payload because the segment header can contain optional fields. Optional fields can be included to match the size of the segments to the route, and to take account of physical limitations of the host machines.

10.5.2.2 Opening, Closing, and Breaking Connections

Logical TCP connections are identified by the IP and port address of the machines at either end. This allows for several TCP connections at a port on a machine since the other end of each connection can be on a different port or machine. Certain port addresses are predefined, such as address 25 that is used by the Simple Mail Transport Protocol (SMTP).

A TCP connection is opened by a transaction of three segments. A machine that wants to open a connection sends an initial segment with the synchronization flag set and an initial value in the sequence number field (see Figure 10.18). The machine that receives it returns a segment that also has the synchronization flag set and contains an initial segment number for the other direction. The returned segment is different from the initial segment because it contains an acknowledgement and indicates the size of the open window. The first machine then sends a segment similar to the second with an acknowledgement and a window size.

Once a machine has sent an initial segment with the synchronization flag set, it starts a timer and returns the connection to the closed state if no acknowledgement is received before the timer reaches zero. If the machine that receives an initial segment does not wish to open a connection, it returns a forced reset. This is a segment with the reset flag set in the segment header. The second and third messages of the transaction that opens a connection are the same to take account of messages crossing during transmission.

When the entire byte stream has been delivered over a TCP connection, the connection is closed by a transaction of four messages (see Figure 10.19). Four messages are used because each machine indicates that it has completed delivery and the other machine acknowledges the indication.

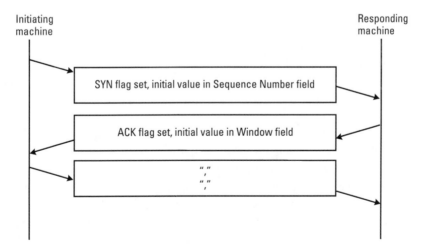

Figure 10.18 Opening a TCP connection.

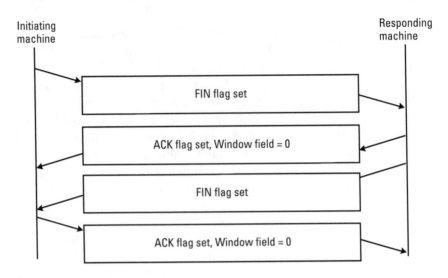

Figure 10.19 Closing a TCP connection.

The first machine to complete its delivery sends a segment with the final flag set. The other machine acknowledges this and closes the window. When the other machine completes delivery then the sequence is repeated in reverse.

A forced reset can be used to actively reject the opening of a connection. It can also be used unilaterally by either end to break a connection before the entire byte stream has been delivered.

10.5.2.3   Urgent Delivery and Fast Delivery

There is a flag that indicates if a segment contains urgent information, such as control information for an application, which must be passed on immediately without regard for the normal octet sequence. There is also an urgent pointer in the header that marks the end of the urgent data in the segment payload.

It can also be important for data in the normal octet sequence to be delivered rapidly, without the usual buffer delays. In this case the sending machine can push information through its own buffer and into a segment, but it also needs to set the push flag in the segment header. The push flag informs the receiving machine that the data should also be pushed through its buffer. The TELNET application protocol uses this mechanism to send control codes so that these can avoid the normal segment queues. TELNET supports a remote user interface, and is in turn used by other application protocols that require a simple remote user interface.

10.5.2.4 Congestion Control and ATM Networks

TCP provides end-to-end congestion control of Internet communications by its flow control mechanism. It does not provide congestion control for the intermediate routers.

ATM could ease congestion on the Internet because it allows new connections to be established between machines. The inverse ATMARP request enables a machine at one end of an ATM connection to discover the ATM address and the IP addresses of the machine at the other end. If a machine knows the ATM address that corresponds to an IP address, then it can create an ATM connection to the machine with the IP address that can bypass busy intermediate routers.

10.6 Routing

When a machine needs to send an IP datagram to a machine on another network or subnet, it is sends the datagram to a router. This router must somehow know where to send the datagram next. Routers keep this information in their routing tables, but the variety of protocols that have been devised toobtain and maintain the information is a testimony to the difficulty of the task.

It is possible to configure the routing tables manually, but this is less effective than an automatic process. Its also does not explain where the information in the routing tables comes from. Calculating the routing information is difficult because there are so many routers and so many links between them, and because the status of the routers and links are constantly changing. Algorithms have been devised to help with this.

10.6.1 Algorithms: Distance Vector and Shortest Path

Two basic algorithms have been devised for calculating routing information. In the distance vector algorithm [14], routers co-operate to calculate the correct routing. The routers advertise their ability to reach networks and give the logical distance to them. This allows other routers to update their own routing tables and advertise the result in turn. The distance vector algorithm is not guaranteed to converge because the calculation is distributed and convergence is disturbed by changes and by the time taken for information about changes to propagate. The algorithm also does not scale well because ultimately every router should advertise its logical distance to every other router.

The other algorithm is the shortest-path-first algorithm. In the shortest-path-first algorithm, each router tests and communicates the status of the links to its immediate neighbors and uses the local status information received from all the other routers to calculate the best path to different networks. The shortest-path-first algorithm converges because it does not involve a distributed calculation and it scales better because less information needs to be communicated. To gain these benefits, each router must calculate all of the routes for itself.

10.6.2 Autonomous Systems, Gateway Protocols, and the Internet Core

The Internet is divided up into autonomous systems of interconnected networks that belong to a single administrative authority such as a business, or a university or government department. It is sensible for only designated routers within an autonomous system to provide routing information for the whole autonomous system to the rest of the Internet because this removes the burden from the other routers. Within an autonomous system, routers are free to communicate as they wish among themselves. In the same way as networks are assigned the network part of their IP addresses, autonomous systems are assigned autonomous system numbers.

Designated routers can use the Exterior Gateway Protocol (EGP) to send routing information for their own autonomous system to routers in other autonomous systems. Within an autonomous system, the proper routing is worked out using an Interior Gateway Protocol (IGP). The EGP can also be used as an IGP within an autonomous system.

The use of the EGP by autonomous systems made the distribution of routing information easier. By itself EGP does not guarantee that a valid route will be found because machine limitations make it impossible for all routes to be analyzed. To overcome the physical limitations or real machines, default routes are used when a specific route is unknown.

Default routing needs to be controlled to prevent destinations becoming unreachable since without proper control a default route for a particular destination could loop back on itself. The National Science Foundation Network (NSFnet) was able to control default routing by acting as the core network for the Internet that did not itself use default routing. Other networks connected to the NSFnet in a tree topology.

The NSFnet architecture grew out of work on ARPAnet (Advanced Research Projects Agency Network) which was created as a test bed for research on networking and which became the backbone of the early Internet. Gateways were established within the ARPAnet to connect to other IP networks. These gateways had complete knowledge of all connected IP routers and exchanged routing information using the Gateway-to-Gateway Protocol (GGP), which is now obsolete.

EGP arose as the original flat architecture evolved into a two level hierarchy with core gateways and exterior gateways to allow the core gateways to communicate with the exterior gateways. At this point the most common IGP was the Routing Information Protocol (RIP). With the creation of the NSFnet backbone and regional networks, the EGP and the various IGPs were used until EGP was replaced by the Border Gateway Protocol (BGP).

The constant pressure from the increasing flow of data needed to support the routing architecture caused a further change to Classless Inter-Domain Routing (CIDR). This involves the replacement of the two-level hierarchy with a generalized multi-level hierarchy and corresponding changes in BGP. Subsequent developments, involving specialized paths and explicitly routed flows are discussed in the final chapters of the book.

Default routing can be misused. A lazy IGP can cause IP datagrams that are destined for itself to be sent over the default route to another autonomous system. The other autonomous system will route the datagrams to the correct destination in the first autonomous system.

Using the NSFnet to control routing is contrary to the spirit of the Internet. The possibility of a router sending datagrams over a default route when it does not care to route them itself is even more contrary to the spirit of the Internet, but it also demonstrates the inventiveness that is so important in a vice.

10.6.3　The Exterior Gateway Protocol (EGP)

An exterior gateway router communicates with other exterior gateway routers in neighboring autonomous systems using the EGP protocol [15]. The EGP protocol supports a distance vector algorithm because it allows routers to tell each other about the distances to the networks that they can reach. A router informs its neighbors of the distances between their common network and networks within its own autonomous system. The EGP update message that is used can either be unsolicited or sent in response to an EGP poll message from another gateway router (see Figure 10.20).

The EGP does not allow a router to calculate the paths to every network for itself because communication outside of the NSFnet is limited to neighbors that share a common network. Routers outside of the NSFnet can only indicate networks within their own autonomous systems.

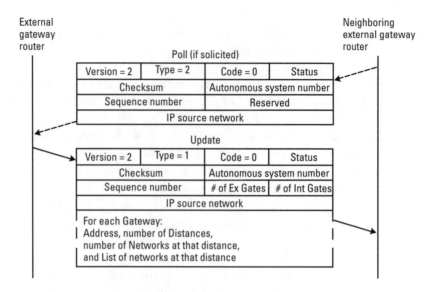

Figure 10.20 Solicited and unsolicited EGP updates.

10.6.3.1 Updates and Polls

An EGP poll message specifies the common reference network with respect to which distances are sought. Like all EGP messages, a poll message starts with a ten octet EGP header that contains six fields and a checksum (see Figure 10.20). This header indicates the autonomous system of the router that sent the message. It also contains a sequence number that allows responses to be matched to requests. The status field in the update message indicates whether or not the update is unsolicited.

The update message specifies a common network and the number of gateway routers on the network that it contains information about. It also specifies if these are interior gateways to within its own autonomous system or exterior gateways to other autonomous systems. For each router it gives the address and a contour image of distances and a list of networks for each distance.

Separate update messages are sent for interior and exterior gateways because the updates do not indicate which information is for which type of gateway. Only a router in the NSFnet should identify reachable networks outside of its own autonomous system.

10.6.3.2 Who Is My Neighbor?

Two routers on the same network that belong to different autonomous systems become neighbors by exchanging EGP neighbor-acquisition messages.

One of the routers sends an acquisition request to the other, which replies with a confirmation or a refusal (see Figure 10.21). Routers stop being neighbors when one sends a cease message to the other, which responses with a cease acknowledgement.

All EGP neighbor-acquisition messages have the same format, which is the common EGP message header followed by two fields. The poll interval field tells the other router what the minimum time must be between its polling requests. The hello interval tells the other router what the minimum time must be between its hello messages. These two fields are only needed for acquisition requests and confirmations because other acquisition messages do not result in neighbors.

A router checks that a neighbor is still reachable by sending hello messages. The neighbor should respond to these with I-H-U (I Heard You) messages. Both of these messages consist of just the EGP header with a type value to indicate that they are EGP neighbor-reachability messages. Hello messages and I-H-U messages have different values in their code fields and the value of their status fields can indicate that the sending router has gone down.

EGP is the historical external gateway protocol. This has been largely superseded by the Border Gateway Protocol (BGP) for exchanging routing information between autonomous systems. Internal Gateway Protocols

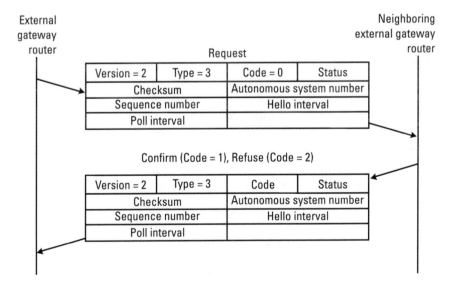

Figure 10.21 Agreeing to be neighbors.

(IGPs) are also needed to distribute routing information within autonomous systems.

10.6.4 The Routing Information Protocol (RIP) and Routing Loop Avoidance

Although EGP was not developed for use within autonomous systems, it can also be used as an IGP that supports a distance vector algorithm. EGP is not well suited for use within an autonomous system because it relies on the existence of the Internet core to prevent routing loops. A better IGP that supports the distance vector algorithm is the Routing Information Protocol (RIP) [16].

RIP limits the maximum distance that a router can use to sixteen hops. Although this in turn limits the size of the autonomous system, it has the property of eliminating routing loops. A loop can be created because of the delay in propagating information about failures. This delay can lead a router to believe that a route that leads back to itself is an alternative that it can use when a valid route fails. Limiting the distance to the destination eventually removes routing loops. This is because each router in the loop tells the router upstream from it that the distance is larger than the distance that the downstream router has told it. As the route length is updated, it grows in size until it is has more than sixteen hops and is removed.

There are a number of techniques that counteract routing loops, but none of these offer a perfect solution. The split-horizon technique reduces destabilizing feedback by prohibiting a router from advertising a route back to the router that first advertised it, but this does not prevent the creation of large routing loops. The hold-down technique reduces destabilizing feedback by prohibiting a router from accepting new and potentially false information about a route to a network until the information that a route is not available has time to propagate. Unfortunately, the hold-down technique also delays the acceptance of valid alternative routes when a route is lost. The poison-reverse technique calls for the distance to be advertised as infinite as a positive indication that a route has been lost, but this needs to be broadcast immediately for the route to be eliminated quickly and this can create an avalanche of broadcasts.

10.6.5 The Border Gateway Protocol (BGP)

Within the original Internet core the HELLO protocol was used as an IGP. This has since been superseded by the Border Gateway Protocol (BGP) [17]. Both support the distance vector algorithm, but the HELLO protocol indicated distance in terms of delay rather than router hops and so routers could react to congestion. Like the RIP, these protocols are limited by the problems of stability and scaling that occur with the distance vector algorithm. The alternative is to use a protocol that supports the shortest-path algorithm.

10.6.6   Open Shortest Path First (OSPF)

The Open Shortest Path First (OSPF) [18] protocol is an IGP that supports the shortest-path algorithm. The OSPF protocol has a number of enhanced features. These include routing by the type of service indicated in the headers of the IP datagrams, partitioning of autonomous systems into areas, load balancing across multiple routes, and authentication to prevent phony routing messages.

Routers that use the OSPF protocol are called OSPF neighbors. OSPF neighbors send hello messages to each other to confirm that they are still reachable. An OSPF hello message includes the addresses of other routers that can be used to contact it and the addresses of other OSPF neighbors that have recently send hello messages to it.

A router keeps the entries in its routing table up to date by sending OSPF link-status requests to its neighbors. New information is sent in OSPF link-status-update messages and these are acknowledged with OSPF link-status-acknowledgement messages.

The OSPF protocol also helps routers to calculate the network topology by enabling neighbors to exchange database information. This is particularly useful for initialization because it allows a new router to obtain a map of the entire autonomous system if it has a single OSPF neighbor. Each entry in an OSPF database-description message consists of the IP address of a particular router and information about one of its links. These links can either be to networks or to other routers. The entire database can be communicated in a sequence of messages.

10.7   Summary

The basic units of IP communications are datagrams that can be carried across a wide variety of networks. IP datagrams have headers that contain addresses consisting of a network part and a local part. The actual addresses used for transport on a network are its link-level addresses. IP addresses are mapped onto link-level addresses using the Address Resolution Protocol (ARP) specific to the network.

An ARP request is broadcast on the local network and contains an IP address. The reply to an ARP request is send by the machine with the matching IP address and returns the corresponding link-level address. The Reverse ARP request asks an address server to return the IP address that corresponds to a contained link-level address.

ARP and RARP are tailored to the different types of networks. The Bootstrap Protocol (BOOTP) and the Dynamic Host Configuration Protocol (DHCP) are application layer protocols that support address resolution independently of the local network. This is possible because IP supports broadcasting that it independent of how broadcasting is performed on specific networks. Multicasting differs from broadcasting because a group of destinations is specified for multicasting. Membership of multicasting groups on the Internet is control by the Internet Group Management Protocol (IGMP).

If the local network is ATM then a different approach to address resolution is needed because most ATM networks often do not support broadcasting or multicasting. The ATM Address Resolution Protocol (ATMARP) allows a machine to ask an ATM address server for the ATM link-level address that corresponds to an IP address. ATMARP is designed to handle ITU-T and ATM Forum link-level addresses and mixed addresses. The ATM equivalent of an Internet network is a Logical IP Subnet (LIS) and it consists of a group of ATM machines that have the same network part in their IP addresses.

The Internet is controlled using the Internet Control Message Protocol (ICMP). ICMP messages are carried in IP datagrams. They are used locally to redirect IP datagrams and to provide machines with a subnet mask that enables them to decide when to send datagrams to a router when a network is partitioned into subnets. ICMP messages also indicate when datagrams are known to have been discarded and allow echo-testing (pinging) of a remote machine and the measurement of relative transmission delays.

IP supports both assured and non-assured communication. Non-assured communication is carried in the payloads of User Datagram Protocol (UDP) datagrams that are in turn carried by IP datagrams. The Simple Network Management Protocol (SNMP) is an application protocol that uses UDP.

Assured communication of a sequence of octets is handled by sending it as Transmission Control Protocol (TCP) segments. Each TCP segment is carried in an IP datagram. TCP provides confirmation of delivery and ensures that the octet sequence is delivered in the correct order and without duplication. Both TCP and UDP use port numbers on machines, but TCP uses the numbers at either end to identify logical TCP connections. TCP also controls the establishment and release of these connections.

Routing of datagrams across the Internet is agreed between autonomous systems that are controlled by different organizations. Traditionally, routing between autonomous systems is coordinated using the Exterior Gateway Protocol (EGP) but this has been superseded by the Border Gateway Protocol (BGP). Routing within an autonomous system is coordinated by an Interior Gateway Protocol (IGP).

Routing protocols support either a distance vector routing algorithm or a shortest-past algorithm. The shortest-path algorithm has technical advantages but requires more calculation. The Open Shortest Path First (OSPF) protocol is a popular IGP that supports the shortest-path algorithm. The EGP and BGP both support the distance vector algorithm.

References

[1]   Comer, D. E., *Internetworking with TCP/IP: Volume 1— Principles, Protocols and Architecture,* New Jersey: Prentice-Hall, 1995.

[2]   Postel, J., *Internet Protocol,* RFC 791 (http://www.ietf.org/rfc/rfc791), September 1981.

[3]   Deering, S., and R. R. Hinden, *Internet Protocol, Version 6 (IPv6) Specification,* RFC 2460, (http://www.ietf.org/rfc/rfc2460), December 1998.

[4]   Mogul, J. C, and Postel, J., *Internet Standard Subnetting Procedure,* RFC 950, (http://www.ietf.org/rfc/rfc950), August 1985.

[5]   Plummer, D. C., *Ethernet Address Resolution Protocol,* RFC 826, (http://www.ietf.org/rfc/rfc826), November 1982.

[6]   Fenner, W., *Internet Group Management Protocol, Version 2,* RFC 2236, (http://www.ietf.org/rfc/rfc2236), November 1997.

[7]   *LANE v2.0 LUNI Interface,* ATM Forum af-lane-0084.000, July 1997.

[8]   *LAN Emulation over ATM Version 2—LNNI Specification,* ATM Forum af-lane-0112.000, February 1999.

[9]   Laubach, M., and Halpern, J., *Classical IP and ARP over ATM,* RFC 2225, (http://www.ietf.org/rfc/rfc2225), April 1998

[10] Bradley, T., Brown, C., and Malis, A., *Inverse Address Resolution Protocol,* RFC 2390, (http://www.ietf.org/rfc/rfc2390), August 1998.

[11]   Postel, J., *Internet Control Message Protocol,*
        RFC 792, (http://www.ietf.org/rfc/rfc792), September 1981.

[12]   Postel, J., *User Datagram Protocol,*
        RFC 768, (http://www.ietf.org/rfc/rfc768), August 1980.

[13]   Postel, J., *Transmission Control Protocol,*
        RFC 793, (http://www.ietf.org/rfc/rfc793), September 1981.

[14]   Baker, F., *Requirements for IP Version 4 Routers,*
        RFC 1812, (http://www.ietf.org/rfc/rfc1812), June 1995.

[15]   Mills, D. L., *Exterior Gateway Protocol Formal Specification,*
        RFC 904, (http://www.ietf.org/rfc/rfc904) April 1984.

[16]   Malkin, G., *RIP Version 2,* RFC 2453, (http://www.ietf.org/rfc/rfc2453), November
        1998.

[17]   Rekhter, Y., and Li, T, *A Border Gateway Protocol 4 (BGP-4),*
        RFC 1771, (http://www.ietf.org/rfc/rfc1771), March 1995.

[18]   Moy, J., *OSPF Version 2,* RFC 2328, (http://www.ietf.org/rfc/rfc2328), April 1998.

11

Internet Applications

But did thee feel the earth move?
  —Ernest Hemmingway

It is the applications of the Internet that has shaken the world, not its fundamental protocols. Internet applications have changed how people work and play. In this chapter we will see how the original trinity of applications evolved into the World-Wide Web, and we will how new applications may be more easily developed.

11.1 Introduction

In the early days of the Internet, there were three primary applications. By the end of the millennium, only one of these remained in widespread use. Remote login to computer systems using TELNET has been largely replaced by remote access via Internet browsers using the HyperText Transfer Protocol (HTTP). Likewise, although the File Transfer Protocol (FTP) remains useful when uploading files to a remote system, downloading is also normally performed using an Internet browser.

The early application that is still popular is electronic mail (e-mail). The initial version of e-mail was extended to allow the attachment of arbitrary files, leading to an explosion in its use. In many organizations, e-mail has replaced telephony as the dominant form of communications.

The Simple Network Management Protocol (SNMP) is also an early Internet application that remains in widespread use. SNMP is used to manage the components of the Internet itself. It is treated separately (see Chapter 12) because management of the Internet is topic in its own right and distinct from the normal use of the Internet.

11.2   User-Friendly Addresses: The Domain Name System

The Domain Name System (DNS) [1, 2, 3] is the basis for the user-friendly addresses that people normally associate with the Internet. The addresses used in datagrams of version four of the Internet Protocol (IPv4) are 32 bits long. These IP addresses are often represented as a sequence of four decimal numbers, one for each octet, separated by periods (e.g. 123.231.012.120). The DNS was devised because these addresses are not user-friendly.

For DNS names to be used instead of IP addresses, two things are needed. First, there must be a way of creating DNS names that correspond to IP addresses. Second, there must be a name resolution procedure that allows machines to determine the IP address from the DNS name.

The authority to create DNS names for IP addresses is hierarchically delegated. The naming hierarchy is partitioned as indicated by the periods (".") in the name. The hierarchical structure of a DNS name does not need to match to the structure of the corresponding IP address, or the physical or topological structure of a network.

The Internet authority has defined the highest level of DNS partitioning to be a set of institutions, organizations and geographical regions. Each of these has a specified label (see Figure 11.1). Every full DNS name has an associated type because the name alone does not indicate whether it corresponds to a domain, a machine, or something else such as a mailbox. A machine that wishes to translate a DNS name into an IP address consults a name server. Name servers have a shallow tree structure that allows high-level domains to keep their mappings on a single server. Name servers also know where to find the mappings that they do not control.

When a machine wishes to find the IP address for a DNS name, it sends a DNS query to a name server, which sends back a DNS response. Queries and responses have the same format (see Figure 11.2). The ques-

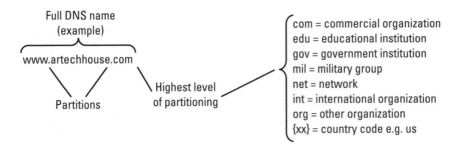

Figure 11.1 Domain names.

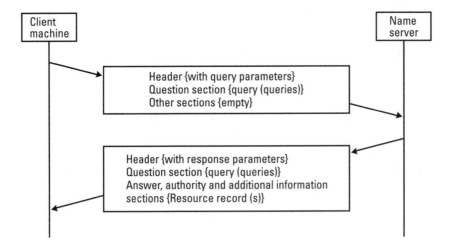

Figure 11.2 DNS queries and responses.

tion section carries the DNS names, and indicates their types since different types may correspond to different IP addresses. The Resource Records (RRs) in the answer section contain the answer to the query. The RRs in the authority section indicate authoritative name servers while the RRs in the additional section give information that may be helpful in using the other records.

If the name server does not know the answer to the query, then it checks to determine whether the client has specified recursive resolution or iterative resolution. For recursive resolution it asks another name server for the answer (see Figure 11.3). For iterative resolution it returns the IP address of the other name server to the client.

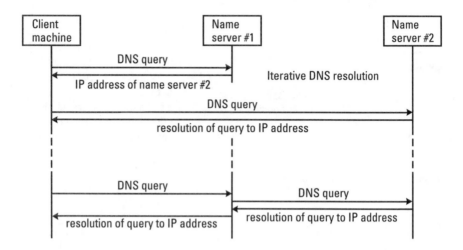

Figure 11.3 Iterative and recursive DNS resolution of unknown mappings.

11.3   The Internet Trinity

The original trinity of Internet applications are TELNET, which allows remote login, FTP, which allows files to be transferred, and e-mail, which allows messages to be sent.

11.3.1 Remote Login: TELNET

TELNET [4] uses TCP to link a user's screen and keyboard to a remote machine. Although the remote machine can be specified by its a domain name, it can also be specified by its IP address to avoid relying on the domain name software. TELNET defines a Network Virtual Terminal (NVT) as the standard interface to a remote system to avoid machine specific variations, such as whether the end of a line of text is denoted by a carriage return or a linefeed.

TELNET treats both ends of the connection in the same way to allow communication between two computers. TELNET also has negotiable options, including one to allow full eight bit characters to be transferred instead of seven bit ASCII characters.

The user's screen and keyboard are attached to a local machine that is the client of the TELNET application (see Figure 11.4). The remote machine is the TENET server. The client machine uses TCP to establish a

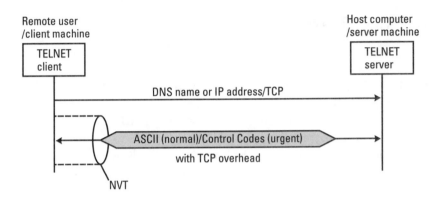

Figure 11.4 TELNET client and server.

connection with the server. Although TCP adds a considerable overhead if only a few bytes are exchanged, this is not noticeable to a human user.

The NVT has standard control functions that are coded as bytes that are preceded with an all ones "interpret-as-command" escape code. For control functions, TELNET uses the TCP urgent data mechanism to bypass the normal TCP segment queue. Control codes allow TELNET options to be negotiated for backward compatibility.

TELNET is also used by other applications such as FTP that need a remote user interface.

11.3.2 The File Transfer Protocol (FTP)

FTP [5] is the basic application that allows machines and human users to transfer files. The FTP client machine can specify the nature of the file. The FTP server machine normally requires the client to provide authorization, although many servers allow anonymous login. The server can also support human users by responding to their help requests.

FTP uses TCP between the client and the server. FTP also uses the basic TELNET protocol without negotiated options for its user interface.

At the start of an FTP session, the remote server uses TELNET to send a request for a name and password back to the user's screen, and the client machine uses TELNET to convey the responses to the server. Once the user is authenticated, the server sends an FTP prompt to the user's screen and responds to the defined FTP commands. These commands include requests from the user to the server to upload, download and display files, and to terminate the session.

11.3.3 Electronic Mail (e-mail)

E-mail [6], in contrast to TELNET and FTP, is so popular with the general public that it was a cornerstone of the 1999 romantic comedy "You've got m@il". Despite its inherent delays, e-mail is more popular than FTP for transferring files because it is more convenient.

E-mail has delays because mail messages are held until they can be transferred to another machine as a background task. This allows e-mail to be delivered even if a remote machine is not immediately available. If repeated attempts to forward mail do not succeed, then the machine tries to send the mail back to the sender.

E-mail is addressed to mailboxes at domain names that look like "mail.box@domain.name.com." A machine that receives mail addressed to one of its mailboxes can interpret the address as the alias for another address or group of addresses and forward the mail accordingly. A machine that forwards mail to a group of addresses is a mail exploder.

The IP application protocol for e-mail is the Simple Mail Transfer Protocol (SMTP). SMTP specifies the control messages that are used to send mail messages from a client machine to a server. The mail messages themselves have body text that is preceded by header text. Both the control messages and the mail messages use readable ASCII text.

To transfer mail, a client must first establish a TCP connection to a server, which replies with an SMTP ready message (see Figure 11.5). The client then sends an SMTP hello message to which the server normally replies with an SMTP OK message. The next set of transactions is derived from the content of the header text in the mail message itself.

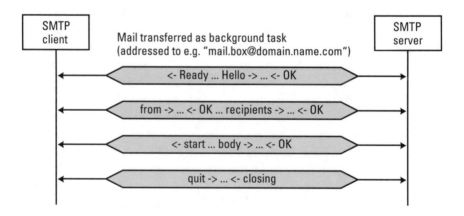

Figure 11.5 SMTP transactions.

The client sends a mail-from message that indicates the origin of the mail and to which error messages should be sent. The server normally replies with another OK message. The client then proceeds to indicate who the recipients are. The server acknowledges each valid recipient with an OK message or indicates if the recipient is invalid. The client can then transfer the body text of the mail message.

The client sends a data message to indicate the start of the body text and sends messages containing the body itself after the server returns a start message. Once the body text has been transferred, the client sends a message consisting of control characters to indicate the end of the body text, and the server acknowledges this with an OK message. The client can then end the session with a quit message, which the server acknowledges with a closing message.

SMTP does not support the transfer of non-ASCII data. To send arbitrary data or files over SMTP it is necessary to use Multipurpose Internet Mail Extensions (MIME) [7].

11.4 Hypertext and the World Wide Web

The dominant Internet application protocol is HTTP [8, 9], the protocol of the World Wide Web (WWW). HTTP gives clients access to servers containing Web pages written in HyperText Markup Language (HTML).

A target Web page is identified by its Uniform Resource Locator (URL), i.e., its HTTP address. The HTTP messages themselves use a format that is similar to e-mail messages with MIME. HTTP also allows access to the same resources as older Internet application protocols, enabling HTTP to supersede these protocols.

Software clients send HTTP requests to servers that return HTTP responses. The requests contain a Universal Resource Identifier (URI), a method, a protocol version identifier, and the MIME-like payload. The method indicates the nature of the request. The response from the server indicates whether or not the request was successful. If the request was successful then the response includes the MIME-like payload. If the request was unsuccessful, the response includes an error code.

HTTP normally operates over TCP. In HTTP version 1.0 (1996) a TCP connection was established and released for each HTTP transaction, but in HTTP version 1.1 (1997) a TCP connection can be used for several HTTP transactions.

11.4.1   HTTP Messages

HTTP messages start with a start-line. The start-line for HTTP requests is a request-line and the start-line for HTTP responses is a status-line. The request-line indicates the version of HTTP, the URI, and the nature of the request - the method (see Figure 11.6). The responding status-line indicates the version of HTTP and includes a numerical status-code that can be processed automatically plus the corresponding reason-phrase that is intelligible to human users.

The start-line is followed by header fields. An empty line is used to indicate the end of the headers and the start of the message-body, the payload of the message. Whether or not the message contains a message-body depends on the particular circumstances. The payload data that is carried in the message-body may be encoded, and the nature of the encoding indicated by a transfer-encoding header.

The transfer-encoding header is a general header that applies to the message as a whole and can be used both in requests and in responses. Headers that apply to the message as a whole, but are specific to requests or responses are called request headers or response headers respectively. Headers that are specific to the payload data are called entity headers.

11.4.2   HTTP Methods

The HTTP methods define the capabilities of HTTP. When a client application wants to retrieve information from a server, it specifies the GET method in the request-line. If the client is a Web browser then the URI to which the GET applies will often be an HTML file.

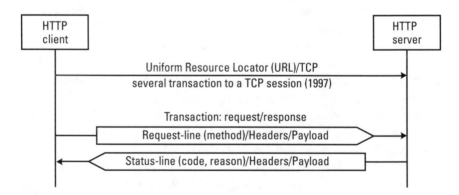

Figure 11.6 HTTP transactions.

The client can use other methods to investigate first. The OPTIONS method allows the client to determine the options and requirements of the resource or the server. The HEAD method allows the client to check the resource by obtaining the headers that would be sent in response to a GET request.

A client that wants to send information to a server specifies the PUT or POST method. The PUT method requests the server to store the entity-body in the request at the URI specified in the request. The POST method requests the server to pass the entity-body to the URI specified in the request.

A client may also specify the DELETE method in a request. The DELETE method does not guarantee that the resource will be deleted since the server may reject it. Even if the server indicates that the DELETE method has been successful, the resource may not have actually been deleted since the response indicates the intention of the server at the time it was sent, and is not a notification of the final result. HTTP diagnostics are supported by the TRACE method.

The GET and HEAD methods are always supported. The server notifies its clients about the methods allowed on resources because these change with time. The supported methods for a resource are indicated in an Allow entity header. The list of methods known to a server are indicated in a Public response header.

11.4.3 Responses to HTTP Requests

If a method is known by a server but not allowed on the specified resource then the status-line in the response should contain a reason phrase of "Method Not Allowed" and the corresponding status-code of 405. If the server does not know the method or does not implement it, then the response should be 501 = Not Implemented.

The 400 series of status-codes are used in the status-line of a response to indicate that the request has been rejected because it is invalid because of a client error. The 500 series indicates that a valid request has been rejected because of a server error. There are three other series of status-codes.

The 200 series indicates a successful request, specifically one that has been received, understood and accepted. Different 200 series codes indicate different outcomes of the request. If the response is simply 200 = OK then there can be other information in the response that depends on the method. If a new resource is created then the response should be 201 = Created. If the request is still being processed then the status-line should indicate 202 =

Accepted. Other responses indicate the amount by which a document being viewed has changed so that the client can request an appropriate update. There is also a code to indicate that header information has been modified, for example by a proxy server.

The 100 series are provisional responses that indicate the initial acceptance of the request or the switch to a new version of HTTP. The 300 series are redirection responses, used to indicate that the client should look elsewhere.

11.5   Remote Procedure Calls

The Remote Procedure Call (RPC) protocol [10] is also based on the client-server paradigm, but the RPC protocol is more flexible than other protocols because the client and server roles are specified for each transaction separately, and not for an entire session. Each call or reply contains generic information such as authorization for the call and parameters that are specific to the call (see Figure 11.7).

The RPC protocol can be used for management transactions and is more versatile and reliable than SNMP because every call is acknowledged and the transactions are not limited to reading and writing at particular addresses. The flexibility in the client and server roles during a session also allows alarms to be sent securely.

RPC can be use either TCP or UDP because there is a reply to each transaction. Care is still needed with UDP because the absence of a reply does not mean that the call has not been received and acted on.

RPC messages use External Data Representation (XDR) [11] in the same way that SNMP uses ASN.1. XDR represents data in a simpler way than ASN.1 because the data does not need to indicate its own type since this is specified in the call or reply and the data does not need to indicate its own length unless it can vary.

XDR variables are defined in declarations that are similar to variable declarations in the C++ programming language. A variable definition is converted into a type definition by prepending the "typedef" keyword to it. A parameter that can be one of a number of different types is defined as discriminated union, a type that is particularly useful in the replies to calls because it allows different information to be returned for successful and unsuccessful calls.

The way in which RPC messages are defined in XDR allows RPC to convey any transactions between machines. It is also capable of supporting

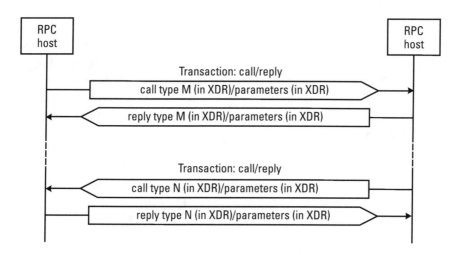

Figure 11.7 RPC transactions.

sophisticated authentication of transactions to protect against threats to security.

11.6 Summary

Of the three early Internet applications, only electronic mail (e-mail) remains in widespread use. Remote login using TELNET and the use of the File Transfer Protocol (FTP) have been largely replaced by Internet browsers using the HyperText Transfer Protocol (HTTP). In addition to these application specific protocols, the Remote Procedure Call (RPC) protocol is designed to support a variety of applications.

Many Internet applications use the Domain Name System (DNS) because IP addresses are not user-friendly. DNS names are partitioned by periods with authority to allocate names assigned through a hierarchy of delegation. Clients that need to determine the IP address corresponding to a DNS name consult a name server. It a name server cannot answer the query then it passes the query on to another server if the client specifies recursive resolution, or returns the IP address of another server if the client specifies iterative resolution.

TELNET can link either a remote user to a host computer or one computer to another, using either a DNS name or an IP address. TELNET uses

TCP, which adds a significant overhead but this is not perceptible to human users. The client acts as a TELNET Network Virtual Terminal (NVT) so that TELNET can be independent of the client hardware. In addition to transporting ASCII characters, TELNET uses an escape code to define control codes that are carried as urgent data in TCP and that allow negotiation of TELNET options. TELNET is used by FTP and other applications that need a simple user interface.

FTP allows machines and human users to transfer files. Once communication is established, the server requests a user name and password from the client. FTP commands for the upload, download and display of files, and the corresponding responses, are conveyed using TELNET.

E-mail is popular and often used to send files because of its convenience, despite the delays caused by machines holding mail until it can be transferred to another machine as a background task. Mail is addressed to mailboxes at DNS names. Mailboxes can themselves be aliases for one or more other e-mail addresses. E-mail is carried by the Simple Mail Transfer Protocol (SMTP). Multipurpose Internet Mail Extensions (MIME) are needed to convey non-ASCII data.

HTTP, which gives clients access to Internet Web sites, has become the dominant IP application protocol. A Web page is identified by a Uniform Resource Locator (URL) that resembles a DNS name and HTTP messages resemble e-mail messages with MIME. HTTP transactions consist of a request and a response. The client retrieves information by a GET method. The client can request information to be stored using the PUT method or simply accepted using the POST method. Servers can indicate which methods they support. Servers also indicate the nature of problems with a request in the status-line of the response.

The RPC protocol differs from other application protocols, both because it is not specific to a particular application and because client and server roles are defined separately for each transaction and not for the entire session. RPC transactions consist of calls and replies that are defined using the External Data Representation (XDR) standard. XDR definitions resemble C++ type definitions.

There is no limit to the potential variety of Internet applications. RPC by itself is a general application protocol that supports any number of specific applications.

References

[1] Mockapetris, P., *Domain Names—Concepts and Facilities,* RFC 1034, (http://www.ietf.org/rfc/rfc1034), November 1987.

[2] Mockapetris, P., *Domain Names—Implementation and Specification,* RFC 1035, (http://www.ietf.org/rfc/rfc1035), November 1987.

[3] Andrews, M., *Negative Caching of DNS Queries (DNS NCACHE),* RFC 2308, (http://www.ietf.org/rfc/rfc2308), March 1998.

[4] Postel, J., and J. Reynolds, *Telnet Protocol Specification,* RFC 854, (http://www.ietf.org/rfc/rfc0854), May 1983.

[5] Postel, J., and J. Reynolds, *File Transfer Protocol (FTP),* RFC 959, (http://www.ietf.org/rfc/rfc0959), October 1985.

[6] Crocker, D.H., *Standard for the Format of ARPA Internet Text Messages* RFC 822, (http://www.ietf.org/rfc/rfc0822), August 1982.

[7] Freed, N., and N. Borenstein, *Multipurpose Internet Mail Extensions (MIME) Part One* RFC 2045, (http://www.ietf.org/rfc/rfc2045), November 1996.

[8] Berners-Lee, T., R. Fielding, and H. Frystyk, *Hypertext Transfer Protocol—HTTP/1.0,* RFC 1945, (http://www.ietf.org/rfc/rfc1945), May 1996.

[9] Fielding, R., et al., *Hypertext Transfer Protocol—HTTP/1.1,* RFC 2616, (http://www.ietf.org/rfc/rfc2616), June 1999.

[10] Srinivasan, R., *Remote Procedure Call Protocol Version 2,* RFC 1831, (http://www.ietf.org/rfc/rfc1831), August 1995.

[11] Srinivasan, R., *XDR: External Data Representation Standard,* RFC 1832, (http://www.ietf.org/rfc/rfc1832), August 1995.

12

Management of the Internet (SNMP)

The secret of managing is to keep the guys who hate you away from the guys who are undecided.

—Casey Stengel

The Simple Network Management Protocol (SNMP) was originally devised for the remote management of IP machines, and routers in particular. Because it is an IP application it can be used across the Internet and is not specific to a particular network. SNMP uses UDP rather than TCP (see Chapter 10) for simplicity so SNMP communication is not assured. SNMP clients manage the Internet and SNMP servers are their remote agents.

SNMP is based on reading (getting) and writing (setting) information in locations in various Management Information Bases (MIBs) that are specific to the managed systems. This allows SNMP to be generic and decouples it from any particular managed system. It also allows SNMP to manage equipment that is not related to the Internet because the detailed information is defined by the MIB. The SNMP's Object Identifier (OID) space supports access both of individual data items and of items that are structured into a table with an index.

Part of the history of SNMP is the conflict between its advocates and the advocates of the Common Management Information Protocol (CMIP). SNMP initially gained acceptance as a measure that was necessary in the interim before the bells and whistles of CMIP were fully forged. Strangely enough, those who developed SNMP implementations were reluctant to discard them in favor of CMIP implementations that had MIBs that were not fully defined. Now that the recriminations have mostly died out, the common phrase that is applied to the situation is "horses for courses". Sadly no one seems particularly willing to admit that there is a course that is unsuitable for his or her own hobbyhorse.

Despite its limitations, SNMP will not be superseded by CMIP, because SNMP is well established and the advocates of CMIP have not been able to get close enough to the guys who are undecided.

12.1 Messages and MIBs

Version 1 [1] is the most widely implemented version of the SNMP. SNMP versions 2 and 3 have been defined (see [2]) but these are not as widely accepted. SNMP uses standard messages with a standard syntax, and the information for each particular managed system is specified in its MIB.

12.1.1 Abstract Syntax Notation One (ASN.1) and SNMP Messages

Both the SNMP messages and the data types of the MIBs are defined using Abstract Syntax Notation One (ASN.1) [3]. MIB data is located in the OID space that is administered by the International Telecommunications Union (ITU) and the International Organization for Standardization (ISO). Four of the five SNMP messages involve gets (reads) and sets (writes) of data at addresses in this space.

There is one SNMP set request, and two types of SNMP get requests with responses to these requests. A simple get request specifies a known MIB location while the get-next request specifies the next highest supported location. The get response returns both the location and the data at that location.

Get-next requests are useful when locations or the index values in tables are unknown. Tables are defined by specifying an OID for a table that consists of rows that contain column entries (see Figure 12.1). A column or set of columns is normally chosen as the index for the table so that OID of the column and the value of the index specify a location in the column.

Set requests are similar to get responses because they include both the MIB location and the data for the location. A get or get-next request is

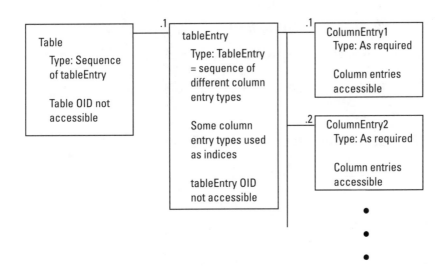

Figure 12.1 Table organization.

needed to read the data if confirmation is required that the set request has been successful. Confirmation is sensible because a set request may be lost as SNMP uses UDP, and because it is not always possible for sets to be acted on immediately.

The final message is the unsolicited trap message that is sent from an agent to a manager. Trap messages are warnings or notifications of alarms that often prompt interrogation of the agent by the manager. Trap notifications include both a generic trap code and a code and data that are specific to a MIB.

Assured communication is not needed for get messages because they prompt a response from the agent. Likewise assured communication is not essential for set messages because these can be confirmed by a subsequent get. Traps are a weakness in SNMP because the lack of assured communications means that neither the manager that should have received a trap message or the agent that has sent it has any indication if the trap message is lost.

SNMP version 1 does not compensate for the loss of trap notifications since it neither provides a means for managers to confirm that a trap has been received, nor provides a means for agents to ask if a trap has been received. An attempt has been made to correct this in the higher versions of SNMP, but it is always sensible for a manager to poll managed devices since they may have become inactive or unreachable.

12.1.2   MIBs and Internet Management

MIB-II [4] is now the accepted basis for the management of the Internet. In addition to its basic groups, it has a set of extension groups (see Figure 12.2) whose OIDs were specified early to minimize subsequent disruption.

12.2   Basic MIB-II Groups and Their Evolution

MIB-II introduced the concept of display strings of printable ASCII characters that are intelligible to human. The Address Translation group in MIB-II is not described because the content of its table has been included in the IP group with complementary information held in the network protocol groups and the group has been deprecated. Subsequent extensions and the Transmission group are treated separately.

12.2.1   The System Group

The system group contains general information about a managed system. It contains displayable text about the name of the managed system, the type and version of its hardware and operating system, and its networking software. It also contains the OID of a supplier's subsystem, and the time since the system was last re-initialized (see Figure 12.3).

The name and details of a human contact for the system were added in MIB-II. The administratively assigned domain name of the system was also added, as were the physical location of the system and its service functions, from physical transmission through bridges and routers to hosts and applications.

12.2.2   The Interfaces Group

The interfaces group contains information about the network interfaces of a managed system. It indicates the number and nature of the interfaces, and provides fault and performance statistics about them. Everything except the number of interfaces is kept in an interface table.

The interface table was extended in MIB-II to allow interfaces to operate below the IP level so that bridges could be supported. The table entry that indicates the types of interface was extended and a new column that provides information specific to the physical media was added.

Subsequent experience in developing media specific MIBs showed that the approach in the interfaces group was still inappropriate for some types of

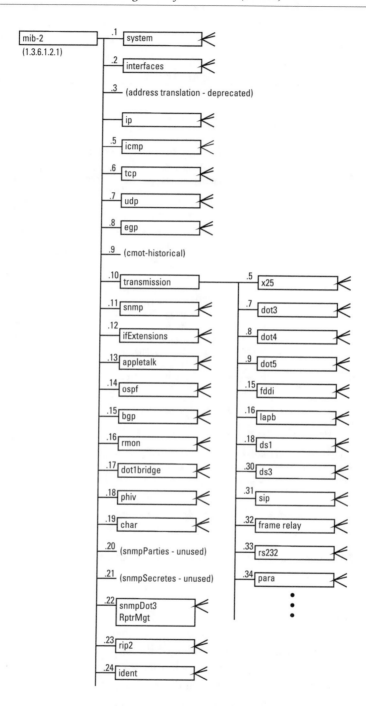

Figure 12.2 Object Identifier (OID) tree for Internet management.

client
manager

agent
server

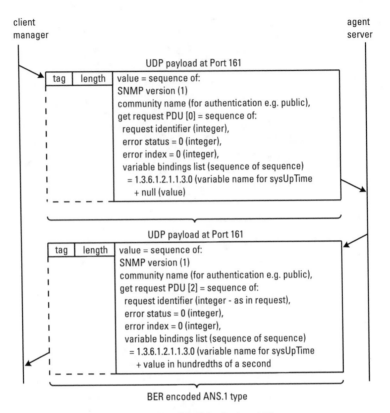

UDP payload at Port 161

| tag | length | value = sequence of: |

SNMP version (1)
community name (for authentication e.g. public),
get request PDU [0] = sequence of:
request identifier (integer),
error status = 0 (integer),
error index = 0 (integer),
variable bindings list (sequence of sequence)
= 1.3.6.1.2.1.1.3.0 (variable name for sysUpTime
+ null (value)

UDP payload at Port 161

| tag | length | value = sequence of: |

SNMP version (1)
community name (for authentication e.g. public),
get request PDU [2] = sequence of:
request identifier (integer - as in request),
error status = 0 (integer),
error index = 0 (integer),
variable bindings list (sequence of sequence)
= 1.3.6.1.2.1.1.3.0 (variable name for sysUpTime
+ value in hundredths of a second

BER encoded ANS.1 type

Note that variable names are "variable OID. 0" for simple variables.

Table variables are identified by sending the OID for the column, which is accessible
and the value or values for the index or indices (e.g. the values of the local and remote
port and IP addresses to get the entry in the connection status column of the TCP
connection table).

If the index values are unknown then column variables may be accesses via a get-next
request on the column and subsequent get-next requests on the returned indexed names

Figure 12.3 Getting the system up time from the system group.

media. In particular there is a need to distinguish between sublayers of the
lower layers and to manage these sublayers independently of the higher layer
protocols that they carry. This challenged the original approach of having a
single row in the interface table for each interface that includes a simple indication of the type of the interface and a reference to a media specific MIB
because this makes it difficult to describe the nature of the sublayers and the
relationship between them.

The solution [5] was to have the rows in the original table describe each sublayer and to add an additional table to describe the relationship between the sublayers. This involved some reinterpretation of the entries in the original table. Two further tables were also added, one with entries that replaced superseded entries in the original table, and another that handles media level addresses and that needs to be interpreted for each specific media.

12.2.3   The Internet Protocol (IP) Group

The purpose of the IP group is to hold statistics about IP datagrams and to hold the machine's routing information and its IP address, or addresses if it is a gateway. It also allows configuration of the machine as a host or as a gateway and of the machine's default time-to-live for datagrams. The IP address table indicates the details and features of the each of the machine's IP addresses and indicates the associated interface. The routing table has configurable entries for each known destination IP address and the details, including metrics, for the route to it.

The group was enhanced [6] to indicate when routes have been lost due to lack of buffer space and to include a table that corresponds to the address translation table in the depreciated Address Translation group. A new column was added to the address table that indicates the size of the largest IP datagram that the associated interface can reassemble. Three columns were also added to the routing table, one that supports subnet masking, one that provides an additional routing metric, and one that is a MIB reference to where additional information on the routing protocol can be found.

12.2.4   The Internet Control Message Protocol (ICMP) Group

The ICMP group has been stable since it was first defined. The purpose of the group is to indicate the monitored performance statistics of the ICMP. The statistics are broken down into the counts of requests and replies for each of the eleven ICMP messages. This breakdown is supplemented by the total counts of ordinary ICMP messages and ICMP error messages sent and received.

12.2.5   The Transmission Control Protocol (TCP) Group

The purpose of the TCP group is to indicate the parameters used in the TCP, to provide statistics about the operation of the TCP, and to monitor the TCP connections. Each row of the TCP connection table identifies the

IP address and port at each end of the connection and indicates the state of the connection. There are also a number of counters, gauges and integer variables that provide information and statistics for the TCP as a whole.

In MIB-II, the entry in the TCP connection table that indicates the state of the connection was extended to include a forced deletion by the managing system. Two additional counters were also added, one to indicate the total number of bad segments received, and one to indicate the number of segments sent with a reset flag.

Subsequent experience resulted in an increase in the size of many of the monitoring records [7].

12.2.6   The User Datagram Protocol (UDP) Group

The UDP group monitors the operation of the UDP protocol. The group has counters to indicate the number of transmitted and received datagrams and the number of received datagrams that could not be delivered. Received datagrams that could not be delivered are broken down into those that had no application at the designated port and those that had some other error.

In MIB-II a table was added with rows that contain the IP addresses and port numbers for UDP on the machine. As for the TCP group, subsequent experience resulted in an increase in the size of counters [8].

12.2.7   The Exterior Gateway Protocol (EGP) Group

Initially the Exterior Gateway Protocol was managed by counting EGP messages at the managed system and keeping track of all the neighbors. Neighbors are recorded in a table that gives their IP address and their status. The number of EGP messages sent and received, with and without errors is noted in separate objects.

MIB-II [4] includes the additional information that experience showed was necessary. For the managed system, the additional information is its autonomous system number. For each neighbor, the additional information is the neighbor's autonomous system number, the breakdown and effects of EGP messages, and the polling mode.

Subsequent development of the management of the EGP was curtailed because EGP was superseded by other gateway protocols. The management of the most significant of these new protocols is defined in extension groups

12.2.8 The Simple Network Management Protocol (SNMP) Group

The SNMP group provides meta-management of the SNMP protocol that is used for management. It is the only group defined in MIB-II [4] for an application protocol. The group allows a management system to enable or disable the use of the authentication trap (see Figure 12.4) and provides statistics on the use of the SNMP to the manager.

12.3 Additional Groups

Additional groups, beyond the basic groups for Internet management, have been also defined, as have groups for the networks that compose it.

12.3.1 Internet Additions

12.3.1.1 TCP User Identification

The purpose of this MIB [9] is to identify the users associated with TCP connections. It consists of a single table that is indexed by the local and remote IP addresses and ports for the connection. The remainder of each row indicates whether user information is available, the nature of the operations system, the user ID plus miscellaneous information. The repertoire of user IDs and miscellaneous information is also given for the connection.

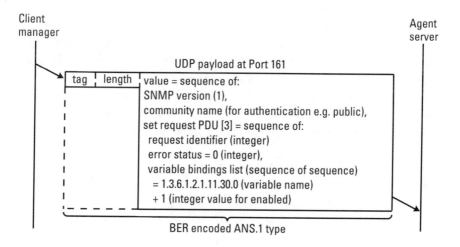

Figure 12.4 Enabling the SNMP authentication trap.

12.3.1.2   Open Shortest Path First (OSPF)

The OSPF routing protocol has a MIB [10] that is extensive and heavily structured. There is an extension for a general group of MIB variables that are global to the OSPF processes, some of which are configurable and some of which are read-only.

There is an area section with tables that indicate OSPF stub areas that the router participates in and their advertised metrics. There is a state section that presents information for debugging. There is a route section that summarizes the known network and routes. There is an interface section that presents the IP interfaces and the service dependent metrics, including virtual interfaces. And there is also a neighbor section that presents the real and virtual neighbors.

12.3.1.3   Border Gateway Protocol (BGP)

The BGP MIB [11] is used to monitor and control BGP implementations. The MIB consists of a small number of system variables, two tables and two specific traps (see Figure 12.5). The system variables indicate the version of the BGP, the local autonomous system number, and the BGP identifier of the local machine. The peer table monitors and controls the state and activity of the connections to BGP peers. The path table presents the raw information received from peers, and some of this information is used to decide on the local routing policy.

12.3.1.4   Routing Information Protocol (RIP)

The RIP-2 MIB [12] is used to monitor and control implementations of version 2 of the RIP. The MIB contains two counters and three tables. The counters are in a global extension in the MIB. They indicate the number of changes that the RIP has made to the routing database and the number of responses sent in return to receive RIP queries. Two of the tables relate to the interfaces. The third table provides information about neighbors that can help in debugging peer relationships. All three tables are in separate MIB branches.

Each row of the interface statistics table is labeled by the IP address of the machine on a subnet. Each row provides information on the number of bad RIP packets received, the number of routes ignored for some reason, and the number of triggered updates sent. There is also a status entry that can be used to delete the row.

The interface configuration table also has rows that are labeled by the IP address of the machine on a subnet. Rows indicate the version or RIP and

Client   Agent
manager   server

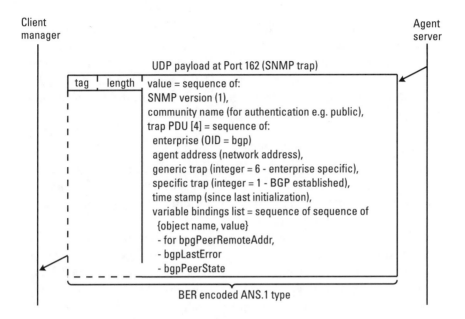

Figure 12.5 Example of the enterprise specific trap for BGP established.

options on the how it is used. Rows also contain authentication information, and can be deleted using the status entry.

Each row in the peer table is labeled by the IP address of a peer RIP machine. Rows indicate when the last update was received from the peer and the version on RIP in the most recently received message. Like rows in the interface statistics table, rows in the peer table also indicate the number of bad packets received and the number of ignored routes.

The revised MIB differs from the original form [13] in two ways. The concept of a routing domain has been dropped, and the concept of unnumbered interfaces has been added. The revised MIB also contains additions that are backwards compatible.

12.3.2 Network Groups

12.3.2.1 Bridges

The MIB [14] for the management of bridges consists of five groups. The first of these is applicable to all type of bridges. The other four groups are particular to various types of bridge and are only implemented if the bridge is of the specified type. They each describe the state of a particular bridge.

There is a separate MIB for source routing bridges [15]. This consists of two groups. One describes the state of the source routing bridge. The other only applies to source routing bridges that support the port-pair multi-port mode.

12.3.2.2   Remote Network Monitoring

The remote network monitoring MIB [16] is used in the interrogation and control of monitoring devices on remote networks. It consists of ten groups. Apart from two history groups that share the same branch, there are separate branches for each of the groups.

The statistics group handles the performance statistics about the interfaces on the remote network. Although the MIB only deals with Ethernet interfaces, the intention is to add other interfaces using the same model. Statistics from past monitoring intervals are kept in the Ethernet history group. The other history group determines the origin of the statistics and the parameters of the sampling.

The alarm and events groups handle trap notifications. The alarm group handles the generation of alarms when thresholds are passed. The event group is used to generate and log other notifications.

The normal hosts group contains statistics about the active hosts on the network. The top hosts group indicates the hosts that are top in the various categories of statistics. Hosts are discovered by monitoring communications on the remote network. Statistics about conversations between pairs of addresses are kept in the matrix group.

There are also two groups that are used to capture packets on the remote network. The filter group is used to select which packets are to be captured, and to control the generation of trap notifications. The packets that have been captured are held in the packets group.

An updated version of the MIB [17] was defined for version 2 of SNMP.

12.3.3   Other Groups

Other groups exist for the management of Ethernet repeaters and for the management of character streams. There are also groups for the management of proprietary networks, in particular DECNet and AppleTalk. No list of groups can be comprehensive because the OIDs space allows for endless evolution.

12.4 The Technology-Specific Groups

The technology-specific groups are continually growing. There is a MIB for Ethernet (IEEE 802.3) and similar CSMA-CD interfaces [18]. There are also MIBs for token bus (IEEE 802.4) [19] and token ring (IEEE 802.5) [20] interfaces, and for Fiber Distributed Data Interchange (FDDI) interfaces [21].

The MIB for ATM [22] is used to manage ATM cross-connects, interfaces and links. It is also used to manage ATM Adaptation Layer 5 (AAL5), which is used to carry IP over ATM. The MIB consists of a number of tables, plus three objects that supports the creation of rows in three of the tables. Other MIBs are also needed to manage a network that uses ATM because ATM, like IP itself, relies on various forms of supporting physical transmission.

There are other groups for the management of transmission services, X.25 communications, standard multiplexing rates for telecommunications, and RS232 and parallel ports. Those for transmission services include Frame Relay and SMDS.

12.5 Summary

Internet management uses the Simple Network Management Protocol (SNMP) that allows locations in an Object Identifier (OID) space to be written to (set) and read from (get). These locations may be individual objects or they may be locations in a table. SNMP also allows the managed system to send trap notifications of events, but there is no guarantee that the loss of a trap message will be detected because, for simplicity, SNMP uses the UDP for end-to-end transport. Versions 2 and 3 of SNMP have been devised to overcome deficiencies in version one, but they are not in widespread use.

There is both a simple get request that specifies the location to be read, including the index if this is a column in a table, and a get-next request that allows the management system to "walk" the Management Information Base (MIB) in the managed system. Both the SNMP messages and the information that they carry are defined using Abstract Syntax Notation One (ASN.1) that has encoding rules that allow them to be placed in UDP payloads.

MIB-II specifies the groups of objects and tables for basic management of the Internet. The System group contains general information. The Interfaces group describes the IP interfaces on a machine. The IP group holds statistics about datagrams, and the ICMP group holds statistics on ICMP messages. The TCP and UDP groups monitor end-to-end transport, while the EGP and SNMP groups monitor these application protocols.

Additional groups are defined for enhancements, such as other routing protocols, and for management of the component networks and network technologies, including ATM and Frame Relay.

References

[1]   Case, J., et al., *A Simple Network Management Protocol (SNMP)*, RFC 1157 (http://www.ietf.org/rfc/rfc1157), May 1990.

[2]   Frye, R., et al., *Coexistence between Version 1, Version 2, and Version 3 of the Internet-standard Network Management Framework*, RFC 2576 (http://www.ietf.org/rfc/rfc2576), March 2000.

[3]   *Information processing systems—Open Systems Interconnection, "Specification of Abstract Syntax Notation One (ASN.1)*, International Organization for Standardization, International Standard 8824, December 1987.

[4]   McCloghrie, K., and M. Rose, *Management Information Base for Network Management of TCP/IP-based Internets: MIB-II*, RFC 1213 (http://www.ietf.org/rfc/rfc1213), March 1991.

[5]   McCloghrie, K., and F. Kastenholz, *The Interfaces Group MIB using SMIv2*, RFC 2233 (http://www.ietf.org/rfc/rfc2233), November 1997.

[6]   McCloghrie, K., *SNMPv2 Management Information Base for the Internet Protocol using SMIv2*, RFC 2011 (http://www.ietf.org/rfc/rfc2011), November 1996.

[7]   McCloghrie, K., *SNMPv2 Management Information Base for the Transmission Control Protocol using SMIv2*, RFC 2012 (http://www.ietf.org/rfc/rfc2012), November 1996.

[8]   McCloghrie, K., *SNMPv2 Management Information Base for the User Datagram Protocol using SMIv2*, RFC 2013 (http://www.ietf.org/rfc/rfc2013), November 1996.

[9]   St. Johns, M., *Identification MIB*, RFC 1414 (http://www.ietf.org/rfc/rfc1414), February 1993.

[10] Baker, F., and R. Coltun, *OSPF Version 2 Management Information Base*, RFC 1850 (http://www.ietf.org/rfc/rfc1850), November 1995.

[11] Willis, S., and J. Burruss, *Definitions of Managed Objects for the Border Gateway Protocol*, RFC 1269 (http://www.ietf.org/rfc/rfc1269), October 1991.

[12] Malkin, G., and F. Baker, *RIP Version 2 MIB Extension*, RFC 1724 (http://www.ietf.org/rfc/rfc1724), November 1994.

[13] Malkin, G., and F. Baker, *RIP Version 2 MIB Extensions*, RFC 1389 (http://www.ietf.org/rfc/rfc1389), January 1993.

[14] Decker, E., et al., *Definitions of Managed Objects for Bridges*, RFC 1493 (http://www.ietf.org/rfc/rfc1493), July 1993.

[15] Decker, E., et al., *Definitions of Managed Objects for Source Routing Bridges*, RFC 1525 (http://www.ietf.org/rfc/rfc1525), September 1993.

[16] Waldbusser, S., *Remote Network Monitoring Management Information Base*, RFC 1757 (http://www.ietf.org/rfc/rfc1757), February 1995.

[17] Waldbusser, S., *Remote Network Monitoring Management Information Base Version 2 using SMIv2*, RFC 2021 (http://www.ietf.org/rfc/rfc2021), January 1997.

[18] Kastenholz, F., *Definitions of Managed Objects for the Ethernet-like Interface Types*, RFC 1643 (http://www.ietf.org/rfc/rfc1643), July 1994.

[19] McCloghrie, K., and R. Fox, *IEEE 802.4 Token Bus MIB*, RFC 1230 (http://www.ietf.org/rfc/rfc1230), May 1991.

[20] McCloghrie, K., and E. Decker, *IEEE 802.5 MIB using SMIv2*, RFC 1748 (http://www.ietf.org/rfc/rfc1748), December 1994.

[21] Case, J., and A. Rijsinghani, *FDDI Management Information Base*, RFC 1512 (http://www.ietf.org/rfc/rfc1512), September 1993.

[22] Tesink, K., *Definitions of Managed Objects for ATM Management*, RFC 2515 (http://www.ietf.org/rfc/rfc2515), February 1999.

13

ADSL Transmission

New lamps for old.
—Aladdin

Asymmetrical Digital Subscriber Line (ADSL) technology allows old ways of using telephone lines to be swapped for new ways that give much higher data rates. In fact ADSL goes one better because narrowband services do not have to be removed when ADSL is introduced. It looks like an offer you can't refuse.

ADSL challenges the Hybrid Fiber Coax (HFC) technology that cable operators can use. HFC supports bi-directional transmission with individual end-users over legacy plant that was designed for broadcast cable television. ADSL in contrast, and later perhaps High-speed Digital Subscriber Line (HDSL) or Very-high-speed Digital Subscriber Line (VDSL) technology, allows higher data rate communication over legacy plant that was designed for bi-directional voice communications with end-users (see Figure 13.1). These other technologies are described in more detail elsewhere (see Chapter 18 and reference [1] of this chapter).

Another metaphor for ADSL is that of putting new wine in old skins, since it is the way that old telephone lines are used that is changed and not the lines themselves. This may serve as a warning. The danger of putting new wine in old skins is that it may ferment and split the skins. Likewise putting

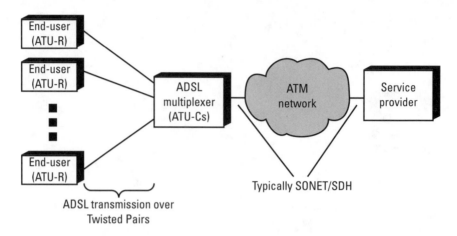

Figure 13.1 Typical ADSL Application

adaptive ADSL transmissions in a cable bundle may cause a reaction like fermentation and produce unpredicted emergent behavior.

Despite the danger of putting new wine in old skins, the quotation from "Aladdin" may be the more apt since ADSL will release the genie of the Internet from the lamp of narrowband access.

13.1 Tones, Modulation, and Coding

ADSL transmission uses a large number of discrete tones that are modulated to carry data and it decreases the sensitivity to noise by Trellis coding.

13.1.1 Tones

Standard ADSL [2] uses Discrete Multi-Tone (DMT) transmission with tones that start at 4.3125 kHz and are spaced 4.3125 kHz apart. There are 255 possible tones downstream, from an ADSL Termination Unit - Central (ATU-C) to an ADSL Termination Unit - Remote (ATU-R), and 31 possible tones upstream. If ADSL operates on the same line as PSTN or narrowband ISDN the lowest frequency tones are not used. For PSTN the reduction in bandwidth is acceptable since PSTN only needs 4 kHz, but it can be easier to treat ISDN as an incompatible service because it needs an order of magnitude more.

The particular tones used, the energy per tone, and the number of bits per tone are agreed during initialization. These values can also be adjusted during normal operation. Echo-cancellation [1] can be used to allow the same tone to be used in both directions. In simpler implementations, unidirectional tones alone are used. This reduces the data rate downstream, because upstream transmission must use the lower frequency tones that are the least attenuated.

13.1.2 Modulation

Each DMT tone is modulated to a point in an amplitude-phase constellation. These points represent the amplitude and phase of the DTMF tone. The cosine direction (see Figure 13.2) represents the component that is in phase. The sine direction represents the component that is 90 degrees out of phase.

All constellations have the same nominal average power, although the actual average power may differ from this because of adaptation during operation. The level of noise at the tone frequency determines the constellation used, because the lower the noise the more points the receiver can detect.

The number of bits carried by a constellation determines its form. Constellations carrying an even number of bits are square. These constellations are generated recursively from constellations carrying two less bits by splitting each point into four (see Figures 13.2 and 13.3).

The constellation carrying three bits is a special case because it has four arms and a rotational symmetry (see Figure 13.2). Constellations carrying a higher number of odd bits are generated from the square constellation carrying one bit less by extending the edges (see Figure 13.3)

These agreed ADSL constellations are not optimal because they do not stack points in a hexagonal array. Like packing fruit in a tray, there can be more points in the same space if they are packed hexagonally rather than

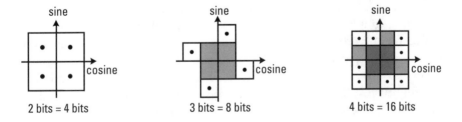

Figure 13.2 Growth of constellations from 2 Bits to 4 Bits.

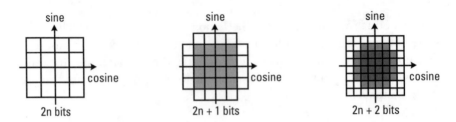

Figure 13.3 Growth of constellations from 4 Bits.

octagonally. The agreed constellations also do not take account of the cost of the points in power, which depends on the distance from the origin.

An optimal constellation has the greatest number of points for a specified average power, with the minimum spacing between points determined by the level of noise. It is more efficient to encode bits across multiple constellations because the number of points in an optimal constellation is rarely a power of two.

13.3.3 Trellis Coding

Trellis coding [3] may be applied before data is mapped onto constellation points to reduce the effect of noise. Trellis coding uses more symbols than needed. The choice of symbol provides additional information about previous data and so reduces the sensitivity to noise.

Trellis coding can have a negligible effect on signal bandwidth if amplitude-phase modulation is used. The cost of Trellis coding is an increase in the complexity of the decoder needed to recover the data. The decoder is more complex because it needs to take account of every subsequent symbol that would be affected by the transmitted data before it makes a decision. Typically the Trellis decoder uses the Viterbi algorithm [3], which eliminates possible sequences of symbols as they become unlikely. Decoding is simplified if dummy data is sent periodically to reset the decoder.

In a simple example of Trellis coding, the value of each bit is represented by one of two possible pairs of orthogonal symbols. The choice of symbol pair depends on the value of the previous bit. The four orthogonal symbols are (+,+), (-,+), (-,-) and (+,-). The second half of the symbol is determined by the current bit, but the first half of the symbol is flipped or not flipped depending on the value of the previous bit.

In this example, if noise causes one half of a symbol to change during transmission then the bit can still be recovered because the information is spread over three half symbols periods (see Figure 13.4). If the four symbols correspond to the four constellation points of phase modulation, the coding gain costs almost no power or bandwidth when the data rate is much less than the carrier frequency.

In general, a Trellis encoder can have a number of states with the new state and the symbol to be transmitted determined by the input data and the previous state. It is called Trellis coding because a trellis pattern is created by the transitions between the states of the encoder (see Figure 13.5).

ADSL data is mapped onto ordered pairs of DMT tones when Trellis coding is used. If there are an odd number of active tones then an inactive tone that carries no bits is used in the first ordered pair. Each active tone carries at least two bits since phase modulation allows two bits to be carried for the same energy as one and the Trellis coding more than compensates for the reduced distance between adjacent points.

Normally the number of constellation points in the second tone of each tone pair is twice what is needed if there was no Trellis coding. The constellation point in the second tone is determined by the least significant bits that are carried by the pair and by the choice made for previous pairs.

Two dummy bits are added to the last two-tone pairs for the ADSL frame to reset the decoder for the next frame. This is done to simplify the decoder as these bits ensure that each transmitted frame has no residual impact on the next frame.

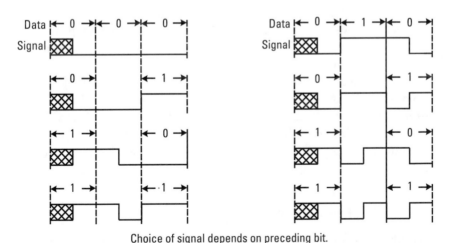

Choice of signal depends on preceding bit.
Signal in subsequent period confirms current bit.

Figure 13.4 A simple example of trellis coding.

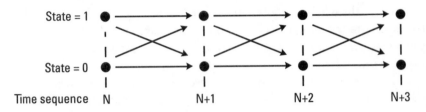

Figure 13.5 Generation of the trellis pattern.

The Trellis coding for ADSL is less than optimal when it is implemented by detecting constellation points and then calculating the transmitted data. This is because information is lost when signals are mapped onto constellation points.

13.2 Frames, Superframes, and Symbols

Tones are ordered according to the number of bits that they carry and then allocated to the transmitted ADSL frames.

13.2.1 Frames, Superframes, and Tone Allocation

ADSL frames are grouped into superframes that are 17 ms long and contain 69 symbol periods (see Figure 13.6). The first 68 of these symbol periods are allocated to the payload frames and the final symbol period is allocated to superframe synchronization.

If 250 tones are used in the downstream direction then payload data would be carried in one million tones every second (see Figure 13.7). In the upstream direction, the data rate is about an order of magnitude less. Tones are ordered according to the number of bits that they carry. The bits at the beginning of the payload frame are assigned to the tones that carry the least bits.

13.2.2 Payload Frames and ADSL Channels

Payload frames contain either the fast channel or the interleaved channel, or both. The interleaved channel has a greater delay because data is held back for transmission on later frames. The number of bytes in a frame and the allocation of bytes to the channels are flexible.

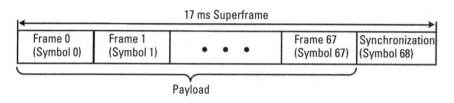

Figure 13.6 Frames, Superframes, and Symbols.

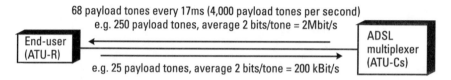

Figure 13.7 Upstream and downstream capacities.

The interleaved channel is allocated to the second half of the frame because tone allocation places the tones that carry the most bits there, and these tones are susceptible to corruption if the transmitted signal is clipped. The interleaved channel is better protected against this corruption because the interleaving makes the forward error correction more effective. Tomlinson precoding [1], which avoids clipping, is not used for ADSL.

13.3   Forward Error Correction

Forward error correction is added to each ADSL channel to reduce the need for retransmission. The interleaved ADSL channel has the more effective forward error correction but the cost is yet further delay.

13.3.1   Error Detection and Error Correction

Error detection and forward error correction work by transmitting the information as a codeword that contains additional information. The redundant information in the codeword allows errors to be detected. In forward error correction there is sufficient redundancy for a limited number of errors to be corrected.

ADSL uses the Reed-Solomon method [3], where the codeword is created by adding additional bytes. These bytes are generated using polynomial division with modulo arithmetic of polynomial coefficients.

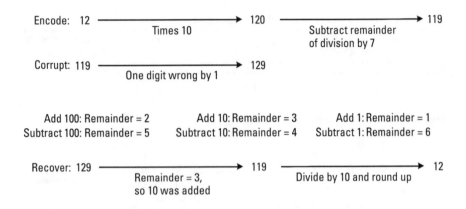

Figure 13.8 Equivalent of Reed-Solomon error correction in simple arithmetic.

The original message bytes are used as the coefficients of a polynomial expression, the message polynomial, and the additional bytes are the remainder when this is divided by the generating polynomial of the Reed-Solomon code. Modulo arithmetic ensures that the coefficients of the remainder are also equivalent to bytes, and these are appended to create the codeword. The generating polynomial is mathematically similar to a prime number in ordinary arithmetic.

In ordinary arithmetic errors can be detected if the only numbers that are used are numbers that are perfectly divisible by another number. For example, numbers that are divisible by seven can be generated by multiplying an initial number by ten and then subtracting the remainder. This allows all errors in arithmetic to be detected except for those errors that introduce multiples of seven.

Errors in arithmetic can also be corrected because different errors produce different remainders when the numbers are divided by seven (see Figure 13.8). There are six ways that a single digit in a three-digit number can be out by one, e.g. a three being written as a two or a four. These six ways correspond exactly to the six possible non-zero remainders when a three-digit number is divided by seven.

This simple example of error correction works because all possible remainders of division by seven are generated by powers of the number ten, which is the base used in ordinary arithmetic. This relationship between the prime number seven and the base ten corresponds to the relationship between the generating polynomial and its length. The effect of noise can be assessed because errors are similar to simple corruption of digits.

13.3.2 Forward Error Correction in ADSL

The bit streams of both the interleaved and the fast channel are scrambled before the Reed-Solomon codewords are generated. For the fast channel a Reed-Solomon codeword is constructed by adding error correction bytes in each frame. Forward error correction in the fast channel is effective if the errors are distributed evenly in time because the chance of overloading a frame with errors is low. Unfortunately errors are often concentrated in bursts that occur in a minority of the frames.

Interleaving the data from different frames makes the error correction more effective because the frames that are error free help to correct the frames that contain many errors. The interleaved transmission also increases the effect of error correction by having a longer Reed-Solomon codeword. Bytes in the interleaved channel are transmitted sequentially but with spaces between them that are filled with the bytes from previous frames. Only the first byte in each frame is not delayed.

13.4 Channels, Ports, and Framing

Payload frames have different modes depending on the number of ADSL channels and whether or not rate adaptation is applied to the bearer channels within them. Rate adaptation is controlled by frame overhead bytes that also perform conventional overhead functions.

13.4.1 Bearer Channels, ADSL Channels, and ATM Ports

Both the fast and the interleaved ADSL channels can support up to seven bearer channels (see Figure 13.9). These form two groups, a simplex (unidirectional) group of four bearer channels (AS0 to AS3) and a duplex (bidirectional) group of three channels (LS0 to LS2). The simplex channels only carry data downstream.

The simplex group normally has a higher data rate than the duplex group because the duplex group is limited by the data rate for upstream transmission. Each bearer channel can operate at a multiple of 32 Kbps. LS0 can also operate at 16 Kbps if it is used to carry signaling for the other bearer channels.

The bearer channels can either be used separately for synchronous transmission or they can be combined to form two ATM ports, one fast and the other interleaved.

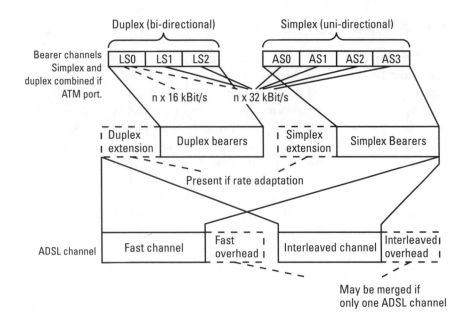

Figure 13.9 Channels, ports, and framing.

13.4.2 Framing and Overheads

The bearer channels are carried within the ADSL channels in one of four possible framing modes. The mode depends on whether the ADSL channels include extension bytes for rate adaptation, whether these extension bytes are used, and whether the ADSL channel overheads are merged if only one channel is present.

In practice this flexibility is not needed because Service Providers obtain greater flexibility at the ATM layer with the simplest ADSL configuration (i.e. one interleaved channel and a single overhead). The only concern is that for delay sensitive applications the fast channel may be necessary instead.

The simplex and duplex bearer channels are multiplexed together onto their ADSL channel and prefixed with an overhead byte. The fast and interleaved ADSL channels carry different information in their overhead bytes and the bytes carry different information in the different frames of the superframe.

The information in the overhead bytes also varies with the framing mode. In the full framing mode, simplex and duplex extension bytes are added after the multiplexed bearer channels. For upstream transmission there is no simplex extension byte because no simplex channels are carried upstream.

The overhead bytes control the use of the extension bytes for rate adaptation by indicating the bearer channels that they are allocated to. They also indicate whether rate adaptation is necessary and whether there is an underflow or an overflow. If there is an overflow then the extension byte is added to the normal bytes. If there is an underflow then the final normal byte of the identified bearer channel is ignored.

The overhead bytes in frame zero of the superframe always carry the CRCs of their ADSL channels. These CRCs are calculated for the previous frame. The overhead bytes also carry 24 indicator bits, which carry alarms and can provide a timing reference, and the Embedded Operations Channel (EOC) and ADSL Overhead control Channel (AOC).

In frames other than 0, 1, 34 and 35, the least significant bit of the fast overhead byte determines whether it controls rate adaptation or whether the overhead carries the EOC instead. In frames other than 0 the interleaved overhead is always allocated to rate adaptation unless the interleaved channel is present but does not carry any bearer channels. If there are interleaved bearer channels and there is no merged overhead then the interleaved extension byte for the duplex bearer channel group is used for AOC when it is not needed to carry an overflow for a bearer channel. The interleaved overhead byte is also used for the AOC if the interleaved channel carries no bearer channels. Unlike the EOC, the AOC is carried in whole bytes.

13.5 Summary

Asymmetrical Digital Subscriber Line (ADSL) transmission uses a very large number of discrete tones that can each be modulated to carry a number of bits. The immunity of these bits to noise can also be increased through Trellis coding, which introduces redundancy to increase the effect that the transmitted data has on the transmitted signal. Tones are ordered according to the number of bits that they carry and are allocated, tones with fewest bits first, to the ADSL frames.

Payload data is carried in 68 of the 69 frames that make up a 17 ms superframe. Payload frames contain either a fast ADSL channel or an interleaved ADSL channel, or both. Forward error correction is added to each ADSL channel payload to reduce the need for retransmission. The interleaved ADSL channel has the greater delay but the more effective forward error correction.

One of four payload-framing modes is used depending on the number of ADSL channels and whether or not there is rate adaptation. Rate adaptation is controlled by the frame overhead bytes, which also carry error detection for the superframes and indication bits for alarms and for network timing. The frame overhead bytes are also used for the Embedded Operations Channel (EOC) and the ADSL Overhead control Channel (AOC).

Each ADSL channel in the downstream direction, from an ADSL Termination Unit-Central (ATU-C) to an ADSL Termination Unit-Remote (ATU-R), can carry a group of bi-directional (duplex) channels and a group of unidirectional (simplex) channels. Only the duplex group of bearer channels can be carried in the upstream direction. The bearer channels on each ADSL channel can be used separately for synchronous channels or combined to create an ATM port.

References

[1]   Gillespie, A., *Access Networks: Technology and V5 Interfacing*, ISBN 089-006-928-X, Norwood, MA: Artech House, 1997.

[2]   *Asymmetrical Digital Subscriber Line (ADSL) Transceiver*, ITU-T Recommendation G.992.1, 1998.

[3]   Clark, G. C., Jr, and Cain, J. B., *Error-Correction Coding for Digital Communications*, ISBN 0-306-40615-2, Plenum Press (New York), 1981.

14

ADSL Management

You can fool all of the people some of the time.
—Abraham Lincoln

Human managers often have a poor idea of what their people are doing and how well they are doing it. Managers in return are not renowned for being profligate with the truth concerning those things that they know. Fortunately, the management of ADSL has not yet reached this level of sophistication.

Operations Systems (OSs) for ADSL must be able to configure the equipment, test the equipment and receive information about its performance and state. To support this, ADSL management specifications have been created that use either the Common Management Information Protocol (CMIP) [1, 2] or the Simple Network Management Protocol (SNMP) [3].

The extent to which these management specifications are used in practice depends on how well they fit into the context of the legacy OSs used by telecommunications operators. As with the quotation for this chapter, you may be fooled if you take the specifications out of context.

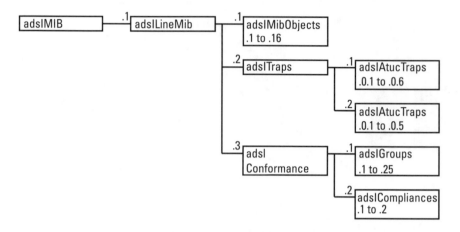

Figure 14.1 Object Identifier tree for the ADSL MIB module.

14.1   The ADSL MIB Module

The SNMP ADSL MIB module complements technology independent SNMP MIB modules to provide a complete basis for the management of ADSL multiplexers and lines. The ADSL module has tables with rows that correspond either to ADSL lines or to their fast and interleaved channels.

14.1.1   Technology-Independent Information

The technology independent physical and logical aspects of a DSLAM are represented in the Entity MIB module in RFC 2037 [4]. The physical components can be racks, shelves, cards or ports. Generic aspects of network interfaces are defined in the Interface MIB module in RFC 2233 [5]. For ADSL these interfaces are either an ADSL line or a fast or interleaved ADSL channel.

Future versions of the SNMP management for ADSL are likely to refer to version two of the Entity module in RFC 2737 [6]. Technology independent aspects of SNMP based management are described in more detail in Chapter 12.

14.1.2   ADSL-Specific Information

The ADSL MIB module [3] has three branches (see Figure 14.1). Information on configuration and status is stored in the first branch (see Figure 14.2). Branch two is used for the traps that generate unsolicited alarms and it is

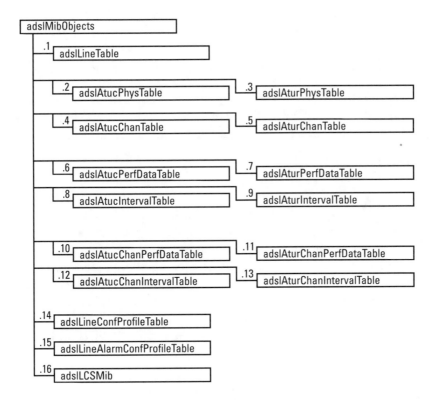

Figure 14.2 Objects in the adslMibObjects branch.

split into a branch for ADSL Termination Unit-Central (ATU-C) alarms and for ADSL Termination Unit-Remote (ATU-R) alarms. The third branch is used in MIB conformance to specify the requirements on implementations.

14.1.2.1 Line and Alarm Configuration

The ADSL line table has five columns and its rows describe specific ADSL lines (see Figure 14.3). Two of the columns describe the coding and type of line. The other columns refer to rows in the line and line alarm profile tables and in a vendor specific table.

Profiles reduce the computer memory needed for the MIB because information that is common to several lines can be stored in a single row of a profile table. The profile tables (see Figure 14.2) have many more columns than the line table. The first column of the profile tables is used to identify rows and the last column allows rows to be created, deleted or modified. Line

	(.1.1)	(.1.2)	(.1.3)	(.1.4)	(.1.5)
	Coding	Type	Specific	ConfProfile	Alarm ConfProfile
One row per ADSL Line	other (1) dmt (2) cap (3) qam (4)	noChannel (1) fastOnly (2) interleavedOnly (3) fastOrInterleaved (4) fastAndInterleaved (5)	(Variable Pointer)	(Snmp Admin String)	(Snmp Admin String)

Figure 14.3 The ADSL line table.

coding and line type could also have been held in as profiles because these are also likely to be the same for a large number of lines.

The line profile table has 30 columns in total (see Figure 14.4). The intermediate columns hold the parameters for ADSL transmission, 14 each for the ATU-C and the ATU-R. These parameters include the fast and interleaved transmission rates, the delay for interleaved transmission, and how spare bandwidth is allocated. They also include the form of rate adaptation, the acceptable range of signal-to-noise ratio, and the interaction between signal-to-noise and adaptation.

14.1.2.2   Alarm Traps and Alarm Configuration Profiles

The alarm profile table has 20 columns (see Figure 14.4). Ten of the intermediate columns hold the parameters for the ATU-C and eight hold the parameters for the ATU-R. One of the ATU-C parameters enables or disables the initialization failure trap, which is the only alarm trap for which no threshold is configured. There is no corresponding ATU-R alarm for initialization failure. The other parameters set the threshold for alarm traps with a value of zero disabling the trap.

Alarms can be triggered by a loss of frame, power or signal quality, by excessive errored seconds, or by changes in channel rates. An alarm can also be triggered by the loss of the link at the ATU-C side, but not at the ATU-R side because communication with the ATU-R is lost when the link is lost and it is sufficient for the ATU-C end to raise the alarm.

14.1.2.3   Physical Transmission and Channel Tables

The physical tables for the ATU-C and ATU-R (see Figure 14.2) contain the same type of information. Each table has eight columns, three for inventory and five for status (see Figure 14.5). The three inventory entries specify the

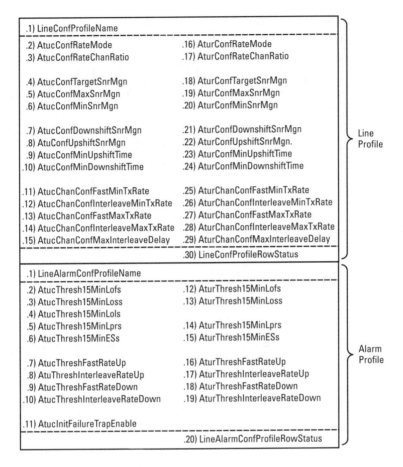

```
.1) LineConfProfileName
.2) AtucConfRateMode                  .16) AturConfRateMode
.3) AtucConfRateChanRatio             .17) AturConfRateChanRatio

.4) AtucConfTargetSnrMgn              .18) AturConfTargetSnrMgn
.5) AtucConfMaxSnrMgn                 .19) AturConfMaxSnrMgn
.6) AtucConfMinSnrMgn                 .20) AturConfMinSnrMgn

.7) AtucConfDownshiftSnrMgn           .21) AturConfDownshiftSnrMgn
.8) AtuConfUpshiftSnrMgn              .22) AturConfUpshiftSnrMgn.
.9) AtucConfMinUpshiftTime            .23) AturConfMinUpshiftTime
.10) AtucConfMinDownshiftTime         .24) AturConfMinDownshiftTime

.11) AtucChanConfFastMinTxRate        .25) AturChanConfFastMinTxRate
.12) AtucChanConfInterleaveMinTxRate  .26) AturChanConfInterleaveMinTxRate
.13) AtucChanConfFastMaxTxRate        .27) AturChanConfFastMaxTxRate
.14) AtucChanConfInterleaveMaxTxRate  .28) AturChanConfInterleaveMaxTxRate
.15) AtucChanConfMaxInterleaveDelay   .29) AturChanConfMaxInterleaveDelay
                                      .30) LineConfProfileRowStatus
```

Line Profile

```
.1) LineAlarmConfProfileName
.2) AtucThresh15MinLofs               .12) AturThresh15MinLofs
.3) AtucThresh15MinLoss               .13) AturThresh15MinLoss
.4) AtucThresh15MinLols
.5) AtucThresh15MinLprs               .14) AturThresh15MinLprs
.6) AtucThresh15MinESs                .15) AturThresh15MinESs

.7) AtucThreshFastRateUp              .16) AturThreshFastRateUp
.8) AtuThreshInterleaveRateUp         .17) AturThreshInterleaveRateUp
.9) AtucThreshFastRateDown            .18) AturThreshFastRateDown
.10) AtucThreshInterleaveRateDown     .19) AturThreshInterleaveRateDown

.11) AtucInitFailureTrapEnable
                                      .20) LineAlarmConfProfileRowStatus
```

Alarm Profile

Figure 14.4 Items in line and alarm profile tables.

vendor, the version number and the serial number. The version number is the same as that used in ADSL initialization.

Four of the five status columns are the same for the ATU-C and the ATU-R. These indicate the measured signal-to-noise ratio, the power transmitted, the calculated attenuation, and the attainable data rate. The current status column is different because more states are possible for the ATU-C than for the ATU-R.

Each row of the ATU-C and ATU-R channel tables (see Figure 14.6) describes a particular fast or interleaved channel. The columns describe the current and previous transmission rates, the size of the data blocks, and delay for interleaved channels.

(.1.1)	(.1.2)	(.1.3)	(.1.4)	(.1.5)	(.1.6)	(.1.7)	(.1.8)
Serial number	Vendor ID	Version number	SnrMgn	Atn	Status	Output pwr	Attainable rate
(Snmp Admin String)	(Snmp Admin String)	(Snmp Admin String)	(tenth dB)	(tenth dB)	noDefect(0) lossOfFraming(1) lossOfSignal(2) lossOfPower(3) lossOfSignalQuality(4)	(tenth dB)	(bps)

One row per ATU, one table per end

ATU-C only { lossOfLink(5) dataInitFailure(6) configInitFaliure(7) protocolInitFailure(8) noPeerAtuPresent(9)

Figure 14.5 The physical ATU-C and ATU-R tables.

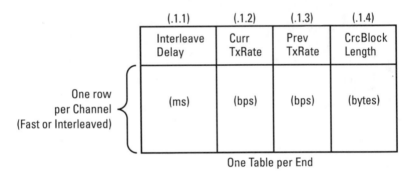

	(.1.1)	(.1.2)	(.1.3)	(.1.4)
	Interleave Delay	Curr TxRate	Prev TxRate	CrcBlock Length
One row per Channel (Fast or Interleaved)	(ms)	(bps)	(bps)	(bytes)

One Table per End

Figure 14.6 The ATU-C and ATU-R channel tables.

14.1.2.4   Performance Monitoring

There are eight tables for performance monitoring, four performance data tables and four interval tables. There are performance-monitoring tables for the physical transmission and ADSL channels for both the ATU-C and the ATU-R. The records in the performance data tables (see Figure 14.7) hold the measures of performance for the total period since there was a reset, for the current 15-minute interval, for the current day, and for the previous day. They also indicate the number of intervals in the corresponding interval table for which data was collected, and the number for which collected data is invalid.

Since reset	Number of intervals	In current 15 min intervals	In current day	In previous day
Loss of .1) Frame, .2) Signal, .3) Link, .4) Power, Number of .5) Err. Secs. .6) Init. Att.	.7) Recorded .8) Invalid	.9) Elapsed secs. Loss of .10) Frame, .11) Signal, .12) Link, .13) Power, Number of .14) Err. secs. .15) Init. Att.	.16) Elapsed secs. Loss of .17) Frame, .18) Signal, .19) Link, .20) Power, Number of .21) Err. secs. .22) Init. Att.	.23) Monit. secs Loss of .24) Frame, .25) Signal, .26) Link, .27) Power, Number of .28) Err. secs. .29) Init. Att.
Number of .1) Rx. Blks. .2) Tx. Blks. .3) Corr. Blks. .4) Unc. Blks.	.5) Recorded .6) Invalid	.7)Elapsed secs. Number of .8)Rx. Blks. .9) Tx. Blks. .10) Corr. Blks. .11) Unc. Blks.	.12)Elapsed secs Number of .13)Rx. Blks. .14) Tx. Blks. .15) Corr. Blks. .16) Unc. Blks.	.17)Elapsed secs. Number of .18)Rx. Blks. .19) Tx. Blks. .20) Corr. Blks. .21) Unc. Blks.

Measures in physical ATU tables, one table for each end, one row per ATU. Numbers refer to the ATU-C table. Shaded measures do not exist in the ATU-R table.

Measures in channel tables, one table for each end, one row per channel (fast or interleaved).

Figure 14.7 Records in performance data tables.

The interval tables (see Figure 14.8) have columns for each of the measures of performance over 15 minute intervals. The rows in the interval tables correspond to the 96 intervals in a 24-hour period. The first column holds the interval number, with an interval number of one indicating the most recent interval. The last column indicates whether or not the row contains valid data.

14.2 The CMIP Model for the Management of ADSL

The CMIP model for ADSL [1, 2] contains similar information to the SNMP ADSL MIB module but it is structured in a different way (see Figure 14.9). The ATU-C and ATU-R are represented by Trail Termination Point (TTP) objects for the ADSL line. When a line TTP object is used at the ATU-C end it points to a profile object that holds similar information to a row in the line profile table in the ADSL MIB module.

Channel TTP objects represent the end-points of ADSL channels within ATU-Rs and ADSL multiplexers, including the adaptation between

Index	Measures	Status
.1) Number	Loss of .2) Frame .3) Signal, .4) Link, .5) Power Number of .6) Err. Secs. .7) Init. Att.	.8) ValidData
.1) Number	Number of .2) Rx. Blks. .3) Tx. Blks. .4) Corr. Blks. .5) Unc. Blks.	.6) ValidData

Measures in physical ATU tables, one table for each
end, one row per ATU.
Numbers refer to the ATU-C table. Shaded measures
do not exist in the ATU-R table.

Measures in channel tables, one table for each
end, one row per channel (fast or interleaved).

Figure 14.8 Interval tables.

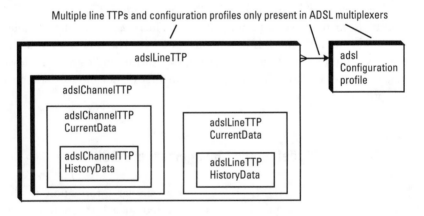

Multiple line TTPs and configuration profiles only present in ADSL multiplexers

Figure 14.9 CMIP modeling of ADSL multiplexers and ATU-Rs.

these end-points and the ADSL transmission. The channel TTP objects are
contained in the line TTP objects.

Ongoing performance monitoring measurements are held in current data objects, different types for ADSL transmission and for channels. These are contained in the relevant line or channel TTP object. The results of past performance monitoring are held in history data objects contained within the corresponding current data objects.

14.2.1 Configuration, Status, and Alarms

Each line TTP object holds information about the line coding, the past and present line rates, and the current line attenuation and signal to noise margin (see Figure 14.10). These objects can also generate alarm notifications to indicate loss of signal, loss or framing, and loss of power.

Each channel TTP object holds information on the type of channel and the past and present channel rates. These objects can also store information on the current length of the Reed-Solomon code used for error-correction and performance monitoring and on the delay introduced if the channel is interleaved.

Although only line TTP objects generate the alarm notifications due to changes in the line conditions, channel TTP objects can generate notifications due to consequent changes in the channel rates. In this case the channel TTP object also holds the size of the change that will cause a notification to be sent.

14.2.2 Profiles

Profiles specify when and if the rate of the line can adapt to local conditions and the maximum power in a spectral window. They also include the target signal to noise ratio and the limits of acceptable signal to noise (see Figure 14.11).

For lines that can change their rates, the profile object holds the signal to noise thresholds and the hysteresis margins for the changes. The profile objects also store the maximum and minimum rates of the fast and interleaved channels, and if both types of channel are supported the profile determines how spare transmission capacity is allocated.

14.2.3 Performance Monitoring

The records that hold the performance measurements use the same structure of 15 minute and one day monitoring periods that are used in the SNMP MIB module. The records of previous measurements in the history data

adslLineTTP	adslChannelTTP	
adslAvailablilityStatus lineCoding supportedChannelTypes supportedOperationalModes currentOperationalMode currentSnrMargin currentAttenuation currentOutputPower currentAttainableRate currentLineRate previousLineRate 1-initFailedNotificationSwitch	channelType currentChannelRate previousChannelRate	Attributes
2-adslConfigurationProfilePointer *2-lineCodeSpecificProfilePointer* *3-allowedOperationalModes*	*1-interleavedMinTxRateAtuC* *1-interleavedMaxTxRateAtuC* *1-maxInterleaveDelayAtuC* *1-interleavedMinTxRateAtuR* *1-interleavedMaxTxRateAtuR* *1-maxInterleaveDelayAtuR* *2-currentCrcBL* *3-upThreshold* *3-downThreshold*	
1-initFailedNotification		Notifications
	3-rateChangeNotification	

Numbers indicate characteristics within the same package and italics indicate conditional characteristics.

Figure 14.10 Characteristics of CMIP ADSL line and channel objects.

objects can only be read. The ongoing measurements in the current data objects can also be reset to zero.

The CMIP line objects can hold up to 10 measures of performance (see Figure 14.12), four more than in the SNMP MIB module (see Figures 14.7 and 14.8). The CMIP model includes additional information about errored seconds and about the success or failure of initialization attempts. The CMIP model also includes an additional measure of the CRC violations detected in the channel.

adslConfigurationProfile		
rateModeAtuC targetSnrMarginAtuC maxSnrMarginAtuC minSnrMarginAtuC rateModeAtuR	targetSnrMarginAtuR maxSnrMarginAtuR minSnrMarginAtuR configuredChannelTypes	Mandatory Attributes
1-downShiftSnrMarginAtuC *1-minDwonShiftTimeAtuC* *1-upShiftSnrMarginAtuC* *1-downShiftSnrMarginAtuR* *1-minDownShiftTimeAtuR* *1-upShiftSnrMarginAtuR* *1-minUpShiftTimeAtuR* *2-fastMinTxRateAtuC* *2-fastMAxTxRateAtuC* *2-fastMin-TXRateAtuR* *2-fastMaxTxRateAtuR*	*3-interleavedMinTxRateAtuC* *3-interleavedMaxTxRateAtuC* *3-maxInterleaveDelayAtuC* *3-interleavedMinTxRateAtuR* *3-interleavedMaxTxRateAtuR* *3-maxInterleaveDelayAtuR* *4-rateChangeRatioAtuC* *4-rateChangeRatioAtuR* *5-lowPowerDataRateAtuC* *5-lowPowerDataRateAtuR*	Conditional Attributes

Number indicate characteristics within the same package and italics
indicate conditional characteristics

Figure 14.11 Characteristics of CMIP profile objects for ADSL lines.

adslLineTTP CurrentData/HistoryData	adslChannelTTP CurrentData/HistoryData	
1-adslLofs *2-adslLols* *3-adslLoss* *4-adslLprs* *5-adslEss* *6-adslSess* *7-adslUass* *8-adslNumFastRetrains* *8-adslFailedFastRetrains* *9-adslFecs*	*1-adslChannelRcvBlocks* *2-adslChannelTxBlocks* *3-adslChannelCorrectedBlocks* *4-adslChannelUncorrectedBlocks* *5-adslChannelCodeViolations*	Conditional Attributes

Numbers indicate characteristics within the same package and italics
indicate conditional characteristics.

Figure 14.12 Characteristics of CMIP performance monitoring objects for ADSL.

14.3 Summary

ADSL management specifications have been created that use either the Common Management Information Protocol (CMIP) or the Simple Network Management Protocol (SNMP). The information that these protocols carry is similar but it is structured in different ways. The SNMP Management Information Base (MIB) uses tables. The corresponding structures in the CMIP model are groups of contained objects of the same type.

Both the CMIP model the SNMP MIB module use profiles to reduce the amount of information that needs to be stored for individual ADSL lines. This information covers the ADSL transmission and the fast and interleaved ADSL channels. Both approaches hold ADSL specific status information that includes the actual transmitted power, the calculated attenuation of the received signal, and the received signal to noise ratio.

Measures of performance are monitored for the current day and for the current 15-minute interval. Records of the results are held for the previous day and for previous 15-minute intervals of the current day. The ADSL MIB module also holds the values of the measures of performance since the managed system was last reset. The measures of performance can generate alarms when certain thresholds are passed.

Alarms can also be generated by other events, such as an initialization failure. The particular alarms generated by the SNMP MIB module and the CMIP model differ in their details, but they include generic alarms for loss of signal and loss of frame. There are also ADSL specific alarms such as indication of a rate change.

The testing of ADSL lines is not covered well in these management specifications.

References

[1]   *CMIP Specification for ADSL Network Element Management*, ADSL Forum Technical Report TR-028, May 1999.

[2]   *ADSL Network Element Management-Part 1: CMIP Model*, ITU-T Recommendation Q.833.1, determined February 2000.

[3]   Bathrick, G., *Definitions of Managed Objects for the ADSL Lines*, RFC 2662 (http://www.ietf.org/rfc/rfc2662), August 1999.

[4]   McCloghrie, K., and A. Bierman, *Entity MIB using SMIv2*, RFC 2037 (http://www.ietf.org/rfc/rfc2037), October 1996.

[5]   McCloghrie, K., and F. Kastenholz, *The Interfaces Group MIB using SMIv2*, RFC 2233 (http://www.ietf.org/rfc/rfc2233), November 1997.

[6]   McCloghrie, K., and A. Bierman, *Entity MIB (Version 2)*, RFC 2737 (http://www.ietf.org/rfc/rfc2737), December 1999.

15

VB5 Access Architecture

A physician can bury his mistakes…
—Frank Lloyd Wright

After their success with the narrowband V5 interface, the experts who developed it realized that they were on to a good thing, so they set out to develop the corresponding broadband interface. With startling audacity they decided to add the letter "B" for broadband and before three further meetings had elapsed they agreed to place the "B" after the "V" rather than after the "5". The VB5 interface had been born.

Like the narrowband interface, momentum developed initially in Europe and then spread out more widely. But unlike the narrowband case, which was specified for 2.048 MBit/s links but not for 1.544 MBit/s, there is no corresponding barrier to the use of the VB5 interface in the USA or Japan. This is because the VB5 interface can be carried over any physical transmission that supports ATM.

15.1 Service Nodes (SNs) and VB5 Interfaces

The VB5 interface [1, 2] is the broadband interface between an Access Network (AN) and a Service Node (SN) (see Figure 15.1). The SN is normally the ATM switch that handles call control signaling from the end-users on the

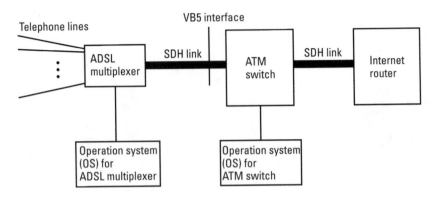

Figure 15.1 Example VB5 application.

AN, but a VB5 interface can also exist if there is no call control signaling. For example, Internet service could be provided to an end-user on an AN that is connected to an Internet router over a semi-permanent ATM network. In this case the Internet router could be the SN.

The AN provides the ATM transmission between end-users and SNs, and any related multiplexing and demultiplexing. The AN can also perform ATM cross-connection and can create bearer connections under the control of the SN.

The VB5 architecture places two constraints on the functionality of ANs. The first is that the AN must not handle call control signaling from end-users, but must pass it to the SN. The second is that any bearer connection that the SN requests the AN to establish can only be between an end-user and a VB5 interface.

It is impractical to provide VB5 interfaces with globally unique labels because there is no central register of VB5 interfaces. Fortunately it is not necessary to have labels which are globally unique because locally unique labels can be defined. VB5 labeling is described in more detail in Chapters 16 and 17.

15.2   Logical Ports and Physical Ports

The VB5 architecture was determined partly by regulatory pressure and partly by the ITU-T concept of an ATM interface. The regulatory pressure insisted that each end-user on an AN could be connected via different VB5 interfaces to different service providers.

The ITU-T, unlike the ATM Forum (ATMF), defines an ATM interface to be a bundle of Virtual Path Connections (VPCs) because it takes the view that an ATM interface may be spread over several physical transmission paths. This approach allows user ports to be connected to ATM switches via ATM cross-connects. Different physical transmission paths may carry ATM connections for the same user port and ATM connections from different user ports may be multiplexed onto the same physical transmission path. Although the ITU-T approach is more architecturally sound, the ATMF approach, which links the physical and logical paths, is simpler and widely accepted.

In the VB5 architecture, bundles of VPCs are called logical ports, and physical ports refer to particular physical transmission paths (see Figure 15.2). VPCs of the same logical port are identified by their VPCIs (Virtual Path Connection Identifiers), which map onto the Virtual Path Identifiers (VPIs) of the physical ports. Different VPIs and physical ports may be used for the same VPCI at different points along the same ATM interface.

The cost of the flexibility of the ITU-T approach is the need for co-ordination of the VPCI labeling and of the mapping of VPCIs to VPIs at the physical ports. In general this aspect of co-coordinated network management is not well specified, although it is covered for the VB5 case (see Chapter 17).

15.2.1 Logical User Ports (LUPs)

A Logical User Port (LUP) consists of those VPCs at a User-Network Interface (UNI) that are associated with a particular VB5 interface. An end-user

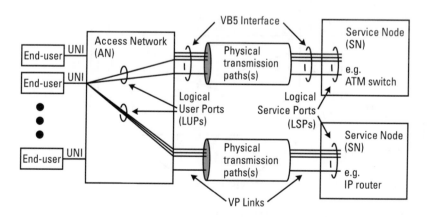

Figure 15.2 Basic VB5 architecture.

may be connected to more than one SN because different VPCs at the UNI can be associated with different VB5 interfaces. However, each VPC at the UNI can only be associated with a single VB5 interface.

There may be one or more physical transmission paths at the UNI, each corresponding to a Physical User Port (PUP). A VPCI, which correspond to particular VPI on a PUP, identifies each VPC of an LUP.

These VPCIs are valid regardless any intermediate VP cross-connection since they are associated with the VPC itself (see Figure 15.3). A VPC on the LUP may terminate in the AN if the Virtual Channels (VCs) within the VPC are individually connected, either semi-permanently or by bearer control signaling from the SN. Semi-permanent connections are configured by the Operations System (OS) of the AN.

A VPC on an LUP that continues on through the AN to the VB5 interface is identified there by its VPCI and LUP number. If a VPC on an LUP terminates in the AN then its LUP number and VPCI do not identify a VPC at the VB5 interface.

15.2.2 The Logical Service Port (LSP)

The Logical Service Port (LSP) is the bundle of VPCs at a VB5 interface. The real-time management co-ordination for these VPCs is handled by the Real-Time Management Co-ordination (RTMC) protocol. The RTMC protocol provides coordination between the AN and the SN. It refers to the

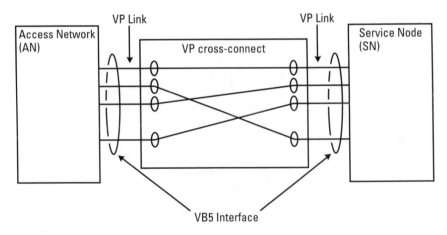

Figure 15.3 VP cross-connection between an Access Network (AN) and a Service Node (SN).

VPs of the VB5 interface in one of two ways, depending on whether or not the VPCs terminate in the AN.

VPCs on the VB5 interface terminate in the AN when the VCs within the VPC are individually connected to an LUP. VPCs also terminate in the AN when their VCs contain signaling which is terminated in the AN. There is always at least one VPC that terminates in the AN because there must be a VPC to carry the VB5 protocols. The VPCs on the VB5 interface that terminate in the AN are identified by the VPCIs of the LSP (see Figure15.4) Each of these VPCIs is mapped onto a VPI of a Physical Service Port (PSP) at the AN. If there is a VP cross-connect between the AN and the SN then these VPCIs are also mapped onto the VPI of a different PSP at the SN.

At the VB5 interface, any VPC that does not terminate in the AN is identified by its LUP number and VPCI. At the UNI this VPCI is mapped onto a VPI of a PUP. At the VB5 interface the LUP number and VPCI are mapped onto a VPI of a PSP.

15.2.3 System Configuration

A VB5 interface will malfunction if its two sides are configured differently. The most obvious mismatch is if the VB5 interface is labeled differently, but this is easy to detect and there is only one label on each side. Each VB5 interface also has about sixteen million possible LUPs.

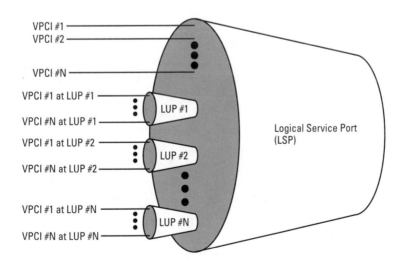

Figure 15.4 Logical User Ports (LUPs) and VPCIs at the Logical Service Port (LSP).

At each LUP, each VPC is identified by its VPCI value, and similarly for the VPCs at the LSP that terminate in the AN. Each VPC has quality of service parameters. In addition, the location of the VCs that carry the VB5 protocols need to be specified together with the parameters that desAt each LUP, each VPC is identified by its VPCI value, and similarly for the VPCs at the LSP that terminate in the AN. Each VPC has quality of service parameters. In addition, the location of the VCs that carry the VB5 protocols need to be specified together with the parameters that describe their ATM adaptation layers.

The complexity of the configuration at either side of a VB5 interface makes the automation of the co-ordination essential to avoid human error. The VB5 management specifications (see Chapter 17) allow this automation to be achieved.

15.3   Signaling and UNI Accesses

There are two types of signaling associated with VB5 interfaces: signaling between the SN and the UNIs; and signaling between the SN and the AN. The signaling between the SN and the UNIs is handled transparently. The signaling between the SN and the AN is fundamental to VB5 interfaces.

15.3.1   The VB5 Protocols

Some of organizations participating in the initial work on VB5 hoped that to avoid defining a new protocol. Their idea was to specify the architecture and to rely on the OAM flows defined in ITU-T recommendation I.610 [3]. Although this did not withstand detailed examination, it led to the two versions of the interface, VB5.1 and VB5.2. The VB5.2 interface, like the narrowband V5.2, supports the dynamic allocation of bearer connections whereas the VB5.1 interface, like the narrowband V5.1, does not.

At an early stage the requirements for real-time and non-real-time co-ordination between the sides of a VB5 interface were differentiated. The difference is due to time required for the OSs on each side to process the information for non-real-time coordination.

The RTMC protocol allows communication between an AN and an SN when information does not need to be processed by an intermediate OS. It allows fast checking of the VB5 interface labels and it supports consistency checking of those VPCs on a VB5 interface that terminate in the AN. It also

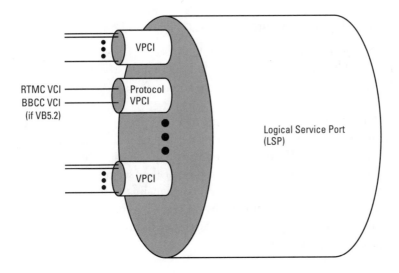

Figure 15.5 VB5 protocol architecture.

allows VPCs and logical ports to be taken in and out of service with the minimum of disruption to active traffic.

The RTMC protocol is present on both VB5.1 and VB5.2 interfaces. VB5.2 interfaces also have the Broadband Bearer Channel Connection (B-BCC) protocol (see Figure 15.5). The B-BCC protocol allows bearer connections on a VB5 interface to be established with LUPs, and it allows these connections to be subsequently modified or deleted. The VB5 protocols are described in more detail in Chapter 16.

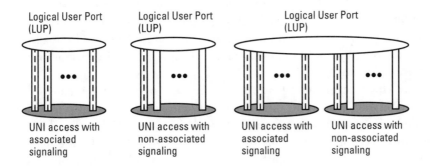

Figure 15.6 UNI accesses within Logical User Ports (LUPs).

15.3.2   User Signaling and UNI Accesses

The SN sees end-users in terms of UNI accesses (see Figure 15.6). Like an LUP, a UNI access consists of a bundle of VPCs between an SN and a UNI, but unlike an LUP, a UNI access need not consist of all VPCs from that UNI.

The distinguishing characteristic of a UNI access is its signaling. This is invisible to the AN because the AN passes user signaling transparently to the SN. In the SN, an LUP consists of one or more UNI accesses.

Each UNI access has a specific type of user signaling. Each VPC in a UNI access can either have its own VC for signaling (associated signaling) or there can be a single VC that handles the signaling for all of the VPCs (non-associated signaling). In either case the signaling protocol is the same for all of the VPCs in the UNI access.

15.4   Comments on the VB5 Architecture

In theory, there can about 16 million VB5 interfaces between an AN and an SN because the interface label has 24 bits. Each interface can also have this number of LUPs. In practice there is no need for this because a single SN does not need an LUP for every member of the population of a thousand planets similar to our own.

Each LUPs can also have about 64 thousand VPCs. Thus a single VB5 interface could support a separate VPC for each of the stars in ten galaxies like our own. As each of these VPCs can in turn support 64 thousand VCCs the potential capacity of even a single VB5 interface is literally astronomical. In practice, broadband ANs only support a few hundred end-users and a handful of ATM connections to each.

This constraint that the only legitimate VB5 connections are from end-users to the VB5 interface invites abuse. User to user connections and VB5 to VB5 connections can be created from legitimate connections because ATM ports can be cross-connected internally within the AN. This is allowed because the physical form of the ports is not specified and an actual physical port need not exist.

The differentiation between real-time and non-real time management co-ordination is artificial and the assumption that OSs need time to process information was a rationalization of the desire to avoid allocating the management co-ordination to the management systems. In reality the management systems need to duplicate much of the co-ordination to allow VB5 interfaces and their RTMC protocols to be brought into operation.

15.5 Summary

A VB5 interface is the broadband interface between an Access Network (AN) and a Service Node (SN). The SN is responsible for the service, which can be that of providing ATM connections on-demand across a broadband network. The AN is responsible for transmission and multiplexing between its end-user and the SN.

There can be more than one VB5 interface between an AN and an SN, but each VB5 interface is a bundle of Virtual Path Connections (VPCs) which are managed together. Both VB5.1 and VB5.2 interfaces have a Real Time Management Co-ordination (RTMC) protocol. VB5.2 interfaces also have a Broadband Bearer Connection Control (B-BCC) protocol.

A Logical Service Port (LSP) is the managed bundle of VPCs at a VB5 interface. The VB5 interface may consist of several physical interfaces, each with its set of Virtual Path and Virtual Channel Identifiers (VPIs and VCIs). Each VPC at the LSP has a Virtual Path Connection Identifier (VPCI), which is supplemented by a Logical User Port (LUP) number if the VPC continues on to an end-user. This allows the LSP to be carried over VP cross-connects between the AN and the SN.

An LUP is the bundle of VPCs at the User-Network Interface (UNI) that are supported by the same SN over the same VB5 interface. Within the SN the LUP may be divided into a number of UNI accesses, each of which use the same signaling protocol, but this is not apparent to the VB5 interface or the AN.

One of the VPCs at the LSP must be reserved for the VB5 protocols. This always contains a Virtual Channel (VC) reserved for the RTMC protocol. For VB5.2 interfaces a second VC is reserved for the B-BCC protocol.

There can be both VP and VC cross-connection within the AN so long as no VPC at any UNI is split between two VB5 interfaces. For VB5.1 interfaces the connections within the AN are controlled by configuration management. For VB5.2 interfaces connections can also be controlled by the SN through the B-BCC protocol, turning the AN into a satellite switch of the SN.

References

[1]   *V-Interfaces at the Service Node (SN) - VB5.1 Reference Point Specification*, ITU-T Recommendation G.967.1, 1998.

[2]   *V-Interfaces at the Service Node (SN) - VB5.2 Reference Point Specification*, ITU-T Recommendation G.967.2, 1998.

[3]   *B-ISDN Operations and Maintenance: Principles and Functions*, ITU-T Recommendation I.610, 1995.

16

VB5 Protocols

Less matter with more art
— Adapted from *Hamlet*

A VB5 interface that connects an access network (AN) to a service node (SN) (see Chapter 15) can take two forms, VB5.1 [1] and VB5.2 [2]. Both forms support error reporting and coordinated testing and maintenance using the Real-Time Management Coordination (RTMC) Protocol, but only VB5.2 interfaces use the Broadband Bearer Connection Control (B-BCC) Protocol that gives the SN control of bearer connections in the AN (see Figure 16.1).

All VB5 interfaces have a virtual channel (VC) within a virtual path (VP) that carries the RTMC. VB5.2 interfaces have a second VC within this VP that carries the B-BCC protocol. No other traffic is carried in this VP.

In *Hamlet* the request was for more matter with less art. This request is not appropriate to the VB5 protocols.

16.1   VB5 Messages and Message Format

Like the narrowband V5 protocols [3, 4], the messages for both VB5 protocols have a common format (see Figure 16.2), which starts with the protocol discriminator. . The coding of the protocol discriminator is the same for all VB5 messages, differentiating them from the messages of other protocols.

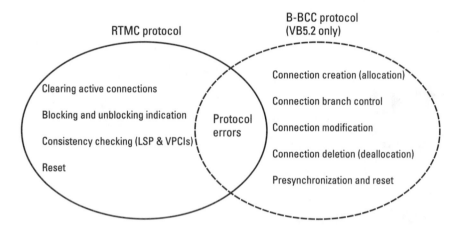

Figure 16.1 VB5 protocol functions.

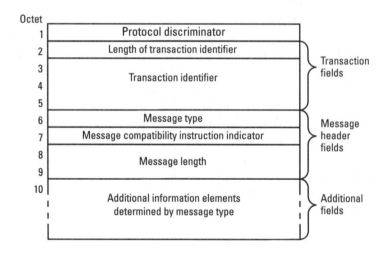

Figure 16.2 VB5 message format.

The remainder of a VB5 message differs from V5 because the common header does not contain a message layer address. Instead, the protocol discriminator is followed by the transaction octets, the message header octets and finally additional octets that depend on the particular message.

The transaction identifier differentiates between different messages of the same type and allows positive or negative acknowledgements to be associated with them. There is a flag in the transaction identifier that is initially set to zero by the side of the interface that originates the transaction and to one by the other side.

Four message header octets follow the four transaction octets. The first of these is the message type octet. The least significant bit of this has a similar function to the transaction identifier flag. The next message header octet is the message compatibility instruction indicator that allows for future evolution.

The remaining two message header octets indicate the number of octets remaining in the message. The message specific information elements and have a structure similar to that of the messages themselves (see Figure 16.3).

16.2 Protocol Errors

There are separate PROTOCOL_ERROR messages for the RTMC and B-BCC protocols, but these have the same content. The messages differ only in the value of the message type field in octet six, and because they are sent in different VCs.

It would have been rational to use a common message number, such as zero, for the protocol error message in both protocols. Instead, two different values of the message type field were allocated to the PROTOCOL_ERROR message in the two different protocols since it was decided not to overlap the message numbers for the two protocols, possibly as an act of worship to the narrowband V5 interface. Non-overlapping values are unnecessary because the protocols are already differentiated from each other as they are carried in different VCs.

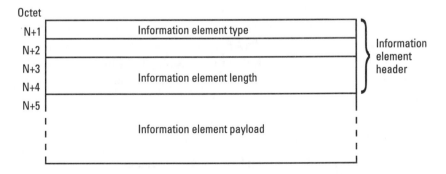

Figure 16.3 Structure of VB5 information elements.

Protocol error messages are sent by either side of the interface in response to inappropriate messages, and they identify the inappropriate messages by using the same value of transaction identifier in their message headers. Each protocol error messages contains a single message specific information element, the protocol error cause.

16.2.1 The Protocol Error Cause Information Element

The protocol error cause information element is only used in protocol error messages. The first octet of its payload contains the protocol error cause value field. The remaining octets contain the diagnostic fields.

If the offending message is identified in the cause value field as being of an unknown type or as not compatible with the state of the process then the diagnostic only needs to include the message type field. If the problem involves the information element then the diagnostic also needs to include the information element identifier field.

One could be tempted to ask why various of the protocol error cause values available for protocol error messages in the narrowband V5 interface are not used here also. The answer is perhaps like God's answer to Job, that it is not possible to understand certain things if you weren't there when the decisions were made.

16.3   The Real-Time Management Coordination (RTMC) Protocol

The RTMC protocol supports error reporting and coordinated testing and maintenance between the AN and the SN. Initially it was hoped that no protocol would be needed for this and that Operations, Administration and Maintenance (OAM) flows defined in ITU-T Recommendation I.610 [5] would support these functions.

Unfortunately, as will shortly be made clear, there are a number of reasons why the I.610 approach does not work. These shortcomings and the refusal of the responsible, or possibly irresponsible, standards groups to consider extending the I.610 OAM flows, lead to the development of the RTMC protocol.

16.3.1 Some Problems with I.610

The I.610 approach to indicating faulty VPs is to mark them with F4 OAM cell flows. Unfortunately, F4 OAM flows cannot be used when VC cross-connection is performed in the AN because the cell flows which would indicate a problem at the User-Network Interface (UNI) terminate in the AN and are not passed on to the SN. But hope, or perhaps just mindless optimism, springs eternal and so it was hoped that if VC cross-connection in the AN was not allowed for VB5.1 interfaces then the I.610 approach could still be used. Unfortunately, it was soon pointed out that this would prevent a common approach to the reporting of faulty VPs for both VB5.1 and VB5.2.

As the requirements for real-time coordination were clarified it also became clear that I.610 OAM flows alone were not sufficient. But it was still hoped that there could be a simplified form of VB5 interface, which would avoid the definition of the RTMC protocol. At this point in the debate, one idea was to define the simplest VB5 interface as that which met those coordination requirements that could be supported by I.610 flows alone. This idea was eventually discarded because of the implications of another ITU-T Recommendation, X.731, which involves the relationship between operational states due to local conditions and administrative states that are determined by the Operations Systems (OSs) controlling the telecommunications equipment.

It emerged that an administrative action performed by an OS on resources used by a VP could prevent the flow of the OAM cells used to mark the VP as faulty. In fact, such an administrative action would create the impression that the VP was available for service since the fault indication would stop. This meant that even the most basic coordination requirements between an AN and an SN could not be guaranteed to be met by OAM cell flows. It also undermined any attempt to seek an increase in the functionality of the OAM cell flows because it is the very existence of the cell flows, rather than their functionality, which cannot be guaranteed.

In the end it was agreed that the definition of an RTMC protocol could not be avoided and the functionality of I.610 was not extended to meet any VB5 requirements.

16.3.2 RTMC Messages

The messages of the RTMC Protocol can be arranged into three sets (see Figure 16.4). One set of messages deals with the blocking of VPCs and the clearing of on-going calls. Another set of messages is used for consistency checking of VPCs. The last set of messages deals with general control functions, in particular VB5 interface ID checking, flagging of detected protocol errors and resetting of VB5 interfaces. All but the PROTOCOL_ERROR messages have both an initial message and a responding message.

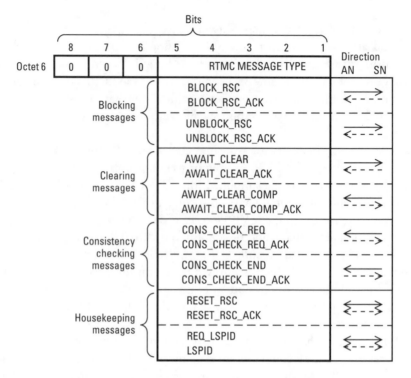

Figure 16.4 RTMC protocol message types.

16.3.2.1 Clearing Active Connections

There are two originating clearing messages and these are used to avoid disturbing active calls when VP connections need to be blocked. Each of the originating messages has an associated acknowledgement message, which indicates whether or not the operation was successful.

When the AN wishes to block VPs without avoid disturbing on-going calls, it needs to send an AWAIT_CLEAR message to the SN because the AN does not know the state of active calls since it does not process call signaling (see Figure 16.5). The SN responds with an AWAIT_CLEAR_ACK message to indicate whether or not it is able to comply with the clearing request. When active calls complete, the SN sends AWAIT_CLEAR_COMP messages and rejects any attempts to set up further calls. The AN responds with AWAIT_CLEAR_COMP_ACK messages as confirmation.

The AWAIT_CLEAR message never refers to a complete logical port. Any resource identifier information elements that it contains must indicate a

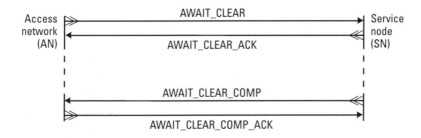

Figure 16.5 Example message flow for RTMC Await Clear messages.

single VP or a range of VPs (see Figure 16.6). If it contains more than one resource identifier information element then it must also contain the repeat indicator information element. It contains no other message specific information elements.

The AWAIT_CLEAR_ACK message contains a result indicator information element saying either that the request was accepted by the SN or that certain resources identified in the request are not known. Unknown resources are identified by resource identifiers specifying either single VPs or ranges of VPs and whether it is the logical port that is unknown or the VPs themselves. If more than one resource identifier is needed to identify the unknown resources then a repeat indicator information element is included. Only these three message specific information elements can be present.

Message	Direction AN SN	Information Elements			
		Rep Ind	B-Rsrc Id	Rsrc Id	Rslt Ind
AWAIT_CLEAR	⟶	1		M, 4	
AWAIT_CLEAR_ACK	◀-----	1		2, 4	M
AWAIT_CLEAR_COMP	⟶	1		M, 4	
AWAIT_CLEAR_COMP_ACK	◀-----				

Key:

RepInd—Repeat indicator   RscId—Resource identifier
B-RsrcId—Blocked resource identifier   RsltInd—Result indicator

M—Mandatory
1—Present if the subsequent information element is repeated.
2—Indicates unknown resources.
4—Indicates VPCI(s) at the LSP or at the LUP.

Figure 16.6 Information elements for RTMC Await Clear messages.

The SN informs the AN about calls completing by identifying the cleared VPs in AWAIT_CLEAR_COMP messages. The VPs are identified either singly or in consecutive groups using resource identifier information elements. For these messages only, the resource identifier also specifies whether the VPs normally carry on-demand calls or whether they are permanently cross-connected in the SN. It is left to the AN to decide when to block the cross-connected VPs since the SN does not process call control signaling for these. The repeat indicator information element must also be included in the message if more than one resource identifier is present. Only these two message specific information elements are present.

Why the AN needs to respond to an AWAIT_CLEAR_COMP message with an AWAIT_CLEAR_COMP_ACK message may be one of life's little unsolved mysteries. The acknowledgement contains no message specific information elements. The message does not appear to do anything that is essential and cannot be easily achieved without it.

16.3.2.2   Blocking and Unblocking

Blocking and unblocking is different for broadband VB5 interfaces than for narrowband V5 interfaces since the narrowband procedure is symmetrical whereas the broadband procedure is asymmetrical. The reason for the difference is that no justification was identified for a real-time indication from the SN to the AN that connections had been blocked or unblocked so simpler asymmetrical procedures were adopted.

A key term here is "real-time". It is clearly necessary for both sides of the VB5 interface to coordinate in taking VPs in and out of service so that both the AN and the SN are aware of the status of connections. However, this coordination is required during installation and reconfiguration, and not during the day-to-day operation of the traffic interface.

VPs or the entire Logical Service Port (LSP) may be blocked either autonomously or after on-going calls have been cleared. The BLOCK_RSC message is used by the AN to inform the SN of faults or administrative constraints on traffic, and in particular if test traffic is permitted (see Figure 16.7). The SN needs to be aware of the inability of the AN to support normal traffic since the SN is responsible for service. The SN responds to a BLOCK_RSC message with a BLOCK_RSC_ACK message.

Likewise, the UNBLOCK_RSC message is used by the AN to inform the SN that normal traffic can again be supported. The SN responds to the UNBLOCK_RSC message by returning an UNBLOCK_RSC_ACK message.

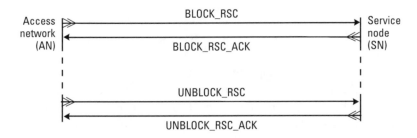

Figure 16.7 Example message flow for RTMC blocking and unblocking messages.

In the BLOCK_RSC message the resources which can be blocked are either single or multiple VPs at a logical port or the entire LSP (see Figure 16.8). Resources are specified in the blocked resource identifier information element. This is the variant of the resource identifier information element with an additional blocking reason indicator field and is only used by BLOCK_RSC messages. If there are several blocked resource identifiers in the message then there must also be a repeat indicator information element. No message specific information elements other than these two should be present.

The BLOCK_RSC_ACK message includes a result indicator information element that is used to specify whether the request was successful or whether certain resources were unknown. Unknown resources are specified

Message	Direction AN SN	Information elements			
		Rep Ind	B-Rsrc Id	Rsrc Id	Rslt Ind
BLOCK_RSC	⟶	1	M, 3		
BLOCK_RSC_ACK	◀-----	1		2, 3	M
UNBLOCK_RSC	⟶	1		M, 3	
UNBLOCK_RSC_ACK	◀-----	1		2, 3	M

Key:

RepInd—Repeat indicator

B-RsrcId—Blocked resource identifier

RscId—Resource identifier

RsltInd—Result indicator

M—Mandatory

1—Present if the subsequent information element is repeated.

2—Indicates unknown resources.

3—Indicates the entire LSP or VPCI(s) at the LSP or at the LUP.

Figure 16.8 Information elements for RTMC blocking and unblocking messages.

in resource identifiers and a repeat indicator is used if resource identifiers are repeated. Only these three types of message specific information elements can be present.

The possibility of blocking the entire LSP is one of those quaint anomalies which add interest to protocol specifications. It is quaint because the message should never be sent, since if the entire LSP is blocked then it cannot carry the message to say that it is blocked. If the message is received then it cannot be true because if it was true then it could not have been received. This possibility could have been excluded by constraining the blocked resource identifier to only refer to single or multiple VPs, as for clearing messages, in which case the anomaly would only arise if the identified VP was that which carries the RTMC protocol.

To be generous we need to allow the BLOCK_RSC message to sometimes mean that the resource will be blocked very soon. If fact we need to be more generous than this because of the acknowledgement message. If the SN really believes that the LSP is blocked, then why does it send an acknowledgement?

The UNBLOCK_RSC message may have a single resource identifier information element or a single repeat indicator information element plus a number of resource identifier information elements. The resource identifiers may specify either the entire LSP or one or more VPs at any of the logical ports. There are no other message specific information elements apart from these two.

The UNBLOCK_RSC_ACK message has a result indicator information element to indicate whether the unblocking is accepted by the SN or if resources are unknown. Unknown resources are specified with one or more resource identifiers with a repeat indicator if there are multiple resource indicators. The resource identifiers specify either the entire LSP or one or more VPs at any of the logical ports. They also specify whether it is the port or the VPs that is unknown. Only these three types of message specific information elements can be present.

There is really no need to have unblocking messages since the blocking reason indicator in the blocking messages has a null value which indicates that there is no longer any blocking. The messages would be simpler if the unblocking messages were deleted and the blocking messages were called status messages. The purpose of the acknowledgement messages is also questionable because the originating messages are notifications, not requests or commands and so they are not something to be complied with, unlike clearing requests. If the resource is unknown then it is more important for this to be raised with the operating systems because it means that there is an inconsistency in the configuration. This is especially important if the unknown resource is the entire LSP.

16.3.2.3 VPCI Consistency Checking

Like blocking, unblocking and clearing, the consistency checking of Virtual Path Connection Identifier (VPCIs) is also an asymmetrical procedure. The procedure is initiated by the SN which first requests a loopback to be applied on a VP in the AN identified by its VPCI (see Figure 16.9). It then sends loopback cells over the connection and checks that they are in fact returned. This confirms that both sides of the interface agree on the connection that is labeled by the VPCI value in the request.

This procedure only applies to VP connections which terminate in the AN because loopbacks can then be applied at the connection end-point. The rationale for this was that a loopback would have to be applied at an intermediate point of a connection if the connection terminates in the customers' equipment since there was no mechanism to request the customer to apply the loopback. This only papers over the cracks since the real problem is the lack of confidence in intermediate loopbacks. This is serious since loopback testing needs to be applied at various points along a connection if it is to be useful for locating faults. Ideally it should be possible to apply a loopback at both the LSP and at the logical user port (LUP) for connections that terminate in customers' equipment since this would assist in locating faults.

A second limitation of the procedure is that it can fail unless the cell labeling and loopback location fields in the loopback cells are used. This is because it is possible that the loopback cells were returned from the wrong place or are not the same cells that were transmitted.

The SN requests the AN to apply a loopback by sending a CONS_CHECK_REQ message. The AN responds with a CONS_

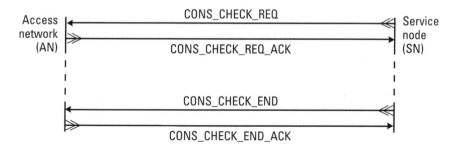

Figure 16.9 Example message flow for RTMC consistency checking messages

CHECK_REQ_ACK message which indicates whether or not the request has been accepted. After sending loopback cells, the SN requests the removal of the loopback by sending a CONS_CHECK_END message. The AN responds to this with a CONS_CHECK_END_ACK message indicating acceptance or rejection of the request to remove the loopback and gives the AN's view of the result of the consistency check.

The CONS_CHECK_REQ message, sent by the SN at the start of the procedure, contains a single resource identifier information element (see Figure 16.10). This is the only message specific information element that is present. The resource identifier has a VPCI field specifying a single VP at the LSP.

The CONS_CHECK_REQ_ACK message, sent by the AN in response, contains a result indicator information element. This informs the SN whether or not the request was accepted. If the request was rejected, it indicates either whether the LSP label of the VP was not recognized or whether the AN had to reject the request without a specific reason. The AN could reject the request if it did not have the necessary resources or if it was already performing a consistency check. If the request is rejected then the resource identifier is returned in the message. No message specific information elements other than these two can be present.

The CONS_CHECK_END messages, which the SN sends to request the removal of the loopback and the result from the AN, contains only a single message specific information element. This is the resource identifier, used to specify the VP at the LSP that is being checked.

Message	Direction AN SN	Information elements			
		Rep Ind	B-Rsrc Id	Rsrc Id	Rslt Ind
CONS_CHECK_REQ	⟶			M, 5	
CONS_CHECK_REQ_ACK	◀-----			2, 5	M
CONS_CHECK_END	⟶			M, 5	
CONS_CHECK_END_ACK	◀-----			2, 5	M

Key:

RepInd—Repeat indicator   Rscld—Resource identifier
B-Rsrcld—Blocked resource identifier   RsltInd—Result indicator

M—Mandatory
1—Present if the subsequent information element is repeated.
2—Indicates unknown resources.
5—Indicates a single VPCI at the LSP.

Figure 16.10 Information elements for RTMC consistency checking messages

The responding CONS_CHECK_END_ACK message contains a result indicator information element, which informs the SN of the successful completion of the consistency check or gives one of four reasons for the check being unsuccessful. In an unsuccessful response, the result indicator may say "unknown", if the LSP label or VPCI in the CONS_CHECK_ END message is not recognized. If the result is not "unknown", then the message may be "rejected", perhaps because no loopback exists. If the result is not "rejected", it can be "not performed", perhaps because the VPCI differs from that on which the loopback was applied. Finally the result may be "failed", perhaps because although the loopback was successfully removed, no cells were found to loop back. If the result is not successful, then a resource identifier information element is included. These are the only two service specific information elements in the message.

The return of the resource identifier in the CONS_CHECK_ REQ_ACK when the request fails is not necessary, because the request is identified by the transaction identifier in the message and the acknowledgement should not return a different value. Similarly, the resource identifiers in the CONS_CHECK_END and CONS_CHECK_END_ACK messages are not necessary if the same transaction identifier is kept for the entire procedure of consistency checking.

16.3.2.4 Verification of the Logical Service Port (LSP) Identifier

Either side of a VB5 interface may check the identity of the other side by sending an REQ_LSPID message (see Figure 16.11). This message is a request to the other side to send back its LSP label. It contains no message specific information elements, so the sending side remains anonymous.

The side of the interface receiving the request responds with an LSPID message. This contains a resource identifier information element specifying

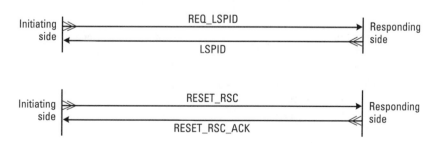

Figure 16.11 Example message flow for RTMC housekeeping messages.

the entire LSP (see Figure 16.12). The label here is the identifier used on the responding side.

No provision is made for the side of the interface receiving the request to reject it. It is assumed that if the message can be received then the LUP identifier has been provisioned.

16.3.2.5   Resetting Resources

Either side of the interface can request either the complete LSP or any VP connection to be reset. The reset causes any related shutting down procedures or VPCI consistency checks to be aborted, and if the reset applies to the complete LSP then the verification of the LSP is also aborted if in progress.

The reset brings the LSP or the VP connections to the unblocked state. These can then be immediately blocked if this is appropriate.

The side requesting the reset sends a RESET_RSC message (see Figure 16.11). This must contain a resource identifier that specifies either the entire LSP or a single or a range of VP connections by logical port (see Figure 6.12). The request must also contain a repeat indicator if the resource identifier is repeated.

The side receiving the request attempts to carry it out and returns a RESET_RSC_ACK message to indicate the result of the attempt. The response must contain a result indicator which can only indicate success or

Message	Direction: AN SN	Information elements			
		Rep Ind	B-Rsrc Id	Rsrc Id	Rslt Ind
REQ_LSPID	←——→				
LSPID	←– –→			M, 6	
RESET_RSC	←——→	1		M, 6	
RESET_RSC_ACK	←– –→	1		2, 6	M

Key:

RepInd—Repeat indicator   RscId—Resource identifier
B-RsrcId—Blocked resource identifier   RsltInd—Result indicator

M—Mandatory
1—Present if the subsequent information element is repeated.
2—Indicates unknown resources.
6—Indicates the LSP.

Figure 16.12 Information elements for RTMC housekeeping messages.

that the resource is unknown. Resources that are unknown are identified using resource identifiers and a repeat indicator if these are repeated.

It would be better if it the reset request could be explicitly applied to an entire LUP since this would avoid the need to specify all the VP connections for an LUP when this is required. The exclusion of the explicit form adds no value. It makes it harder to perform the reset, but does not prevent it.

16.3.3 RTMC Information Elements

The messages of the RTMC protocol are constructed from information elements. Some of these are the generic information elements that were described in Section 16.1. The others are information elements that are specific to the RTMC protocol.

16.3.3.1 The Repeat Indicator

The repeat indicator information element has only single payload octet after the four octets of its header. This octet has a fixed value with other values being reserved for use in the future. The information element only occurs in resource-related messages where its presence indicates that resource information elements may be repeated.

It is difficult to find a better example of redundancy in a protocol than the Repeat Indicator information element. This takes five octets to convey one bit of information that is not actually needed since the message length field indicates this.

16.3.3.2 The Resource Identifier

The resource identifier information element has four, six or eight payload octets depending on whether or not it has VPCI fields. There may be no VPCI fields or one or two VPCI fields, each field being two octets long. If there are no VPCI fields then the complete port is identified. One field is used to identify a particular VP at the port and two fields are used to identify a range of VPs. If present, the VPCI fields follow the logical port identifier field, which is three octets long and identifies the LSP or an LUP. The logical port identifier field is in the second, third and fourth octets of the resource identifier payload.

The first octet of the resource identifier payload contains the resource indicator field. This determines whether the logical port identifier field refers to the service port or a user port. It also determines whether the complete port or particular VPs at the port are identified. Only the complete LSP may be identified. Complete LUPs may not be identified.

The unknown resource indicator field and the cross-connect (XC) flag are also present in the first octet of the resource identifier payload. The XC flag is only meaningful in the await clear complete message where it is used to indicate whether the resource is cross-connected in the SN or whether it is used for on-demand connections. The unknown resource identifier field is only meaningful in acknowledgement messages that indicate that the operation has failed due to an unknown resource. In this case the unknown resource identifier field indicates whether it is the logical port or the VP or VPs at the port that is unknown.

16.3.3.3   The Blocked Resource Identifier

The blocked resource identifier information element has the same payload octets as the resource identifier information element, but these are preceded by a single octet, which contains the blocking reason indicator field. The XC flag and the unknown resource fields from the resource identifier are not meaningful because the information element is not used in await clear complete messages or other responding or acknowledgement messages. It is only used in the initiating messages for blocking.

The blocking reason indicator field is split into an error reason field and an administrative reason field. The error reason field is used to indicate either faults or a failure caused by the non-availability of supporting resources. It has the value of error or no error. The administrative reason field indicates what administrative constrains have been placed on the resource. It has two values to indicate whether all cell flow has been stopped or whether test calls are allowed. It also has a value to indicate that there is no administrative constraint.

16.3.3.4   The Result Indicator

The result indicator information element has just a single payload octet after the four octets of its header. This octet contains a four-bit result indicator field that is used to inform the originating side of the interface whether or not the operation requested by a resource-related initiating message was successful. If the operation was not successful, this information element attempts to indicate why it failed. Resource identifier information element may also be included in the message to identify the resources associated with the failure.

16.3.3.5   RTMC Information Element Anomalies

Resource identifiers are awkward when they are used to specify an unknown LUP because it is necessary to also specify one or more VPs at the port and then say that it is the port rather than the VPs that are unknown.

Resource identifiers are also unnecessarily pedantic when referring to the LSP because they always give its label. It is important to identify a specific LUP because there are several of these, but it is not necessary to identify the LSP because there is only one.

The repeat indicator, as it is defined, adds no significant value to the specification because it does not indicate how often a resource identifier is repeated. This has to be determined from the detailed examination of the message since it cannot be determined from the message length field in the header because resource identifiers vary in length and number. Since the message must be examined in detail in any case to check for consistency, it would be simpler to delete the repeat indicator from the protocol.

16.4   The Broadband Bearer Connection Control (B-BCC) Protocol

The VB5.2 interface extends the VB5 functionality by adding dynamic allocation of broadband connections to the functions supported by the RTMC protocol. It does this in a similar way to the narrowband V5.2 interface by adding an additional protocol, the B-BCC protocol. The B-BCC protocol turns the AN into a satellite switch controlled by the SN.

16.4.1   B-BCC Messages

Most of the messages of the B-BCC protocol are defined in groups of three (see Figure 16.13). The exceptions to this rule are the messages for the deallocation of broadband bearer connections and for the reporting of faults in the AN, which come in groups of two, and the message used for reporting a protocol error, which is independent.

The reason for the groups of three is that request messages are sent from the SN to the AN and acceptance or rejection messages are returned by the AN to the SN. The deallocation request is exceptional because the AN is not allowed to reject it. However, for all of these messages the SN is the

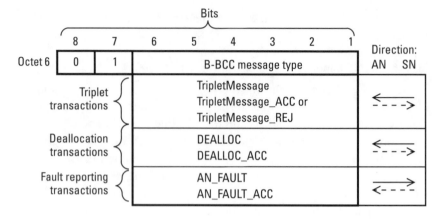

Key:

TripletMessage is AllocationMessage, ModificationMessage, BranchMessage, or
     HousekeepingMessage

AllocationMessage is ALLOC or ALLOC_COMP

ModificationMessage is MODIFY, MODIFY_ABORT, or MODIFY_COMP

BranchMessage is ADD_BRANCH, UPDATE_BRANCH, or DROP_BRANCH

HousekeepingMessage is BBCC_RESET or BBCC_PRESYNC

Figure 16.13 B-BCC protocol message types

master for the connection control since the AN is not responsible for service and does not handle user signaling.

The other messages are for error reporting. The AN reports faults to the SN, which sends an acceptance message to indicate that the report has been received. If either side of the interface detects a protocol error in a received message then it reports it to the side which sent it, but the side which sent the faulty message does not acknowledge the protocol error message in case each attempt to acknowledge a protocol error message also contains an error and causes another protocol error message to be returned.

Unlike the BCC protocol for the narrowband V5 interface, the B-BCC protocol for the VB5 interface has no auditing messages, so auditing of connections must be carried out as part of the higher level coordination between the AN and the SN. Also, unlike V5, two groups of three messages are used for housekeeping messages, the reset and the presynchronization messages.

16.4.1.1 B-BCC Transactions and Negotiations

All B-BCC messages are part of a two-message transaction consisting of a request and a response. For deallocation and AN fault transactions the request message is the DEALLOC request or the AN_FAULT notification, and the normal responses are the DEALLOC_ACC or AN_FAULT_ACC acknowledgements.

The other requests have a choice of a positive or negative response so that the normal transaction is either Message plus Message_ACC, for positive acceptance, or Message plus Message_REJ for negative rejections (see Figure 16.14). In exceptional circumstances the response to any message will be the protocol error message for the B-BCC protocol.

Negotiation requires the transactions themselves to work in pairs with a positive response in the first transaction leading to a subsequent transaction. The initial allocation of a bearer connection and any subsequent modification of its traffic descriptor are negotiated, and the addition of a branch to a point-to-multipoint connection may be negotiated.

In a negotiation, the AN must specify its preference when it accepts a request from the SN since there can be no negotiation if the SN determines parameters unilaterally.

16.4.1.2 Allocation Negotiation (Allocation and Allocation Complete)

The allocation negotiation consists of an allocation transaction and a subsequent completion transaction. The allocation transaction begins with an allocation request (ALLOC) send from the SN to the AN, prompted by incoming or outgoing call or by a set of connections that the SN has been configured to establish using the B-BCC but that are not the result of normal call control.

The allocation request may be for a point-to-point connection or for the root and first branch of a point-to-multipoint connection. In either case the allocation request contains a connection reference number and for point-to-multipoint connections the allocation request also includes a branch identifier.

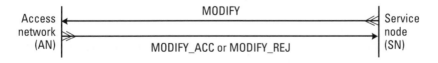

Figure 16.14 Example message flow for B-BCC triplet transactions.

The allocation request also specifies the locations at the user port and service port for the connection but allows negotiation of these. In return the AN must specify the precise VCs for the connection at both ports in the acceptance message (ALLOC_ACC) if these are not those preferred by the SN.

There can be no negotiation of the broadband bearer capability, the quality of service parameter, or the OAM traffic descriptor because these cannot be returned when the AN accepts an allocation request. The last of these three parameters does not have to be specified if there is no OAM traffic, but in all three cases the AN is only informed of the parameters in the allocation request and can only reject the request if they are not acceptable.

The allocation negotiation may be aborted by the AN with a rejection message at either the first step (ALLOC_REJ) or at the second step (ALLOC_COMP_REJ). If the SN wished to abort the negotiation it can either refuse to talk to the AN or issue a deallocation (DEALLOC) request.

The specification does not clearly distinguish between information elements which are negotiated and which are confirmed. Any information element that is returned in an acceptance of an allocation request could be treated as containing a negotiated parameter, but not all of these are confirmed by information elements in the allocation complete (ALLOC_COMP) message. On the other hand, it is foolish to treat information elements in the acceptance of an allocation message as confirmations since it is not necessary to confirm items that are not changed.

Even worse, the specification does not ensure that there is a conclusion to the negotiation since parameters that might not have been agreed during the allocation transaction can be excluded from further negotiation in the completion transaction. This can be seen for the service port connection identifier where the AN may propose a VC to the SN, but the SN cannot confirm or modify this as it can with other parameters.

A better approach would be to specify clearly which parameters may be negotiated in the allocation request, which parameters the AN may change without further negotiation when it accepts a request, and which parameters must be determined in the completion request from the SN. This would be further improved by excluding parameters that can no longer be changed from subsequent messages.

16.4.1.3   Modification Negotiation (Modify, Modify Abort, and Modify Complete)

The modification negotiation allows the ATM traffic descriptor for a point-to-point connection to be re-negotiated. It differs from the allocation negotiation because it is much simpler and because the SN can explicitly abort the negotiation without resorting to a deallocation transaction.

The negotiation starts with the modification transaction and ends with the abort or completion transaction for the modification. The modification transaction starts with a modification request (MODIFY) being sent from the service mode to the AN. All modification requests include a connection reference number to identify the connection, but no branch identification number because modification requests do not apply to point-to-multipoint connections and if they did they would apply to all branches.

The modification request also includes the proposed ATM traffic descriptor and either an alternative or a minimum ATM traffic descriptor as the basis for negotiation. The request can be rejected (MODIFY_REJ) with a cause given if the AN cannot support negotiation within the basis proposed, or the request can be accepted (MODIFY_ACC). In either case an automatic congestion control information element can be included in the response.

In the acceptance an ATM traffic descriptor is included which indicates the best which the AN can support. This can be the initial proposal or the alternative or a descriptor between the initial proposal and the minimum in the request. The acceptance also returns the alternative or the minimum if this differs from best that is already identified.

The SN is then free to abort the modification by sending an abort request (MODIFY_ABORT) to the AN. However, it would be more usual for the SN to send a completion request (MODIFY_COMP) for the negotiation. The completion request includes an ATM traffic descriptor if this is different from the one originally proposed. If neither an abort nor a completion request is received within the allocated time then the AN sends a fault notification (AN_FAULT) to the SN.

16.4.1.4 Branch Transaction and Negotiation (Adding and Updating)

Adding a branch to an existing point-to-multipoint connection can either be done through a negotiation or through a simple transaction. In either case the first step is the same, an add branch transaction.

The add branch transaction begins with an add branch (ADD_BRANCH) request. This request identifies the root by its connection reference number and contains a branch identifier for the new branch. A VC at the user port can be identified explicitly with a user port connection identifier. Alternatively, only the user port or a VP at the port may be identified in the port connection identifier and alternative VPs at the port may be identified with an alternative user port VPCI information element.

The request can be rejected (ADD_BRANC_REJ) with a cause given if the AN cannot support any new branch within the basis for negotiation in the request, or the request can be accepted (ADD_BRANCH_ACC). In either case an automatic congestion control information element can be included in the response.

In the acceptance the AN can simply accept the VC proposed by the SN, in which case the acceptance does not identify a VC and the addition of the branch is completed as a transaction without negotiation. Alternatively the AN can suggest a VC within the basis for negotiation in the request and the second transaction of the negotiation can begin.

If the SN wishes to accept the VC identified either explicitly in the acceptance or by tacit agreement with its initial request, then it sends an update branch (UPDATE_BRANCH) request to the AN. This update branch request identifies the root connection and the branch under negotiation and confirms the user port connection identifier.

The update branch request can be rejected (UPDATE_BRANC_ REJ) with a cause given or the request can be accepted (UPDATE_ BRANCH_ ACC). No automatic congestion control information element is included in either of these responses.

If the SN wishes to abort the negotiation it can either refuse to talk to the AN about it or issue a drop branch (DROP_BRANCH) request.

16.4.1.5   The Drop Branch Transaction

The drop branch transaction supports the release of a branch of a point-to-multipoint connection in the AN. The drop branch (DROP_BRANCH) request identifies the root connection with a connection reference number and the branch(es) to be dropped with a branch identifier list or a branch identifier. A drop branch cannot be used to delete the last branch of a point-to-multipoint connection because no connection would then exist, so for this case a deallocation transaction must be used instead.

Under normal conditions the AN releases the branches and returns an acceptance (DROP_BRANCH_ACC) message to the SN. If the connection number is unknown or if there would be no branches left then a rejection (DROP_BRANCH_REJ) with an appropriate cause value is returned instead. In either case an automatic congestion control information element may be included in the response.

If there are unknown branches in the drop branch request then only the known branches are released and the acceptance is returned. No

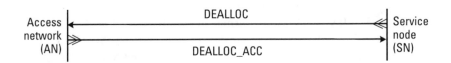

Figure 16.15 Example message flow for BBCC deallocation messages.

indication of the unknown branches is passed to the SN and this needs to be handled by the higher layers of management for the interface.

16.4.1.6   The Deallocation Transaction

The deallocation transaction enables the release of both point-to-point and point-to-multipoint connections in the AN (see Figure 16.15). When it is used for point-to-multipoint connections then the root and all of the branches are released. The deallocation request (DEALLOC) identifies connections to be released with a connection reference number or a connection reference number list.

If there are unknown connections in the deallocation request then only the known connections are released and the acceptance (DEALLOC_ACC) is returned. The only message specific information element that may be present in the acceptance is an automatic congestion control information element. The deallocation request cannot be rejected, but has no effect on connections if it does not refer to any known connections.

No indication of unknown connections is passed to the SN and this needs to be handled by the higher layers of management for the interface.

16.4.1.7   The Housekeeping Transactions (Reset and Presynchronization)

Reset transactions are used during initial start-up of the B-BCC protocol and may be used during restarts after the protocol has become inactive. Presynchronization transactions are used during restarts to determine is a reset is required. In both cases the Signaling ATM Adaptation Layer (SAAL) must be established before the transaction is used.

During start-up a reset request (BBCC_RESET) is sent from the SN to the AN. The AN responds either with an acceptance (BBCC_RESET_ACC) or a rejection (BBCC_RESET_REJ). There are no message specific information elements in the acceptance and only a reject cause in the rejection.

During a restart a presynchronization request (BBCC_PRESYNC) is sent to the AN. If connections have been preserved during the inactive phase and no reset is necessary then the AN responds with an acceptance

(BBCC_PRESYNC_ACC) and normal operation of the B-BCC protocol can continue. If the AN responds with a rejection (BBCC_PRESYNC_REJ) then a reset transaction is initiated and connections need to be synchronized between the AN and the SN.

16.4.1.8   The AN Fault Reporting Transaction

The B-BCC protocol has two mechanisms to report problems. Faults in the AN are reported to the SN, and either side of the interface can inform the other of protocol errors.

If the AN becomes aware of a fault that affects a bearer connection then it informs the SN by sending an AN_FAULT message (see Figure 16.16). Each AN_FAULT message must contain either a connection reference number or a user or service port connection identifier, and it may contain more than one of these.

If a particular branch of a point-to-multipoint must be identified and the connection reference number is used then it has to be supplemented by a branch identifier. If the connection is identified by a user or service port connection identifier then the particular VCC at the port must be indicated.

When the SN receives an AN_FAULT message it responds by sending an AN_FAULT_ACC message back to the AN. This is an unnecessary acknowledgement since messages are already acknowledged at layer two of the protocol, and this message conveys no additional information.

The definition of the AN_FAULT message could be improved because it allows a connection to be identified in three ways simultaneously and does not take account of the implications of this. In particular, there is no need to insist that the branch identifier is included if the branch is already identified by the user port connection identifier. In implementations it sensible to only use one of the three ways in any particular message since this is simple and avoids inconsistency.

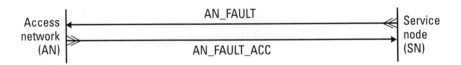

Figure 16.16 Example message flow for AN Fault messages

16.4.2 B-BCC Information Elements

The messages of the B-BCC protocol, like those of the RTMC protocol, are also constructed from information elements. The generic VB5 information elements are described in Section 16.1. Some of the other information elements, particularly those relating to ATM traffic descriptors and quality of service parameters are borrowed from ITU-T recommendations on broadband signaling (see Chapter 8). The information elements described here are those that are particular to the B-BCC protocol.

16.4.2.1 The Connection Reference Number

The connection reference number has three octets in its payload and these form the connection reference number value field. The values of this field uniquely identify individual bearer connections.

16.4.2.2 The Connection Reference Number List

The payload of the connection reference number list contains one or more sets of three octets. Each set of three octets forms a connection reference number value field as used in the connection reference number information element.

The connection reference number list identifies one or more bearer connections. This information element is only used for deallocation requests.

16.4.2.3 The Service Port Connection Identifier

The service port connection identifier has one, three or five octets in its payload, depending on whether it refers to the LSP itself, or to a specific VP, or to a specific VC within a VP. The first payload octet contains the resource indicator field and the connection identifier field. If there are more payload octets then octets two and three contain the VPCI filed and octets four and five contain the VCI filed.

The resource indicator field implies the length of the service port connection identifier because it has three values. It has a value to indicate the entire LSP, in which case there are no VPCI or VCI fields, and it has two other values. These values indicate a specific VP or a specific VC within a VP, implying the presence of the VPCI field or of both the VPCI and the VP fields.

The connection identifier field is ignored unless the message is an allocation request (ALLOC). For allocation requests, the connection identifier specifies either that the combination of VPCI and VCI fields is non-negotiable or only preferred.

16.4.2.4   The User Port Connection Identifier

The user port connection identifier is similar to the service port connection identifier, but it always has three more octets in its payload and it also has a direction field in its first payload octet, in addition to the resource indicator and the connection identifier fields.

The meaning of the resource indicator field is slightly different in the user port connection identifier because it refers to an LUP, rather than the LSP that it refers to in the service port connection identifier. The reason for the three additional payload octets is that it is necessary to specify the particular LUP and these are not needed for the service port because there is only one. The LUP is identified in payload octets two to four.

In payload octet one the direction flag has two values that are used to indicate whether the user port is the originating side or the terminating side of the bearer connection. Payload octets five to six and seven to eight contain the VPCI and VCI fields respectively.

16.4.2.5   Alternative Service and User Port VPCIs

The alternative service port VPCI and the alternative user port VPCI may each be used in allocation requests to specify one or more alternative VPCI at the service or user ports for the VC connection being established. In each case the payload contains one or more sets of two octets and each set of two octets forms a VPCI field.

The two information elements differ only in values of their information element type fields.

16.4.2.6   The Automatic Congestion Level

The automatic congestion level information element has a single octet in its payload and this is used to specify the degree of congestion. The definition of the two thresholds for congestion is not specified explicitly but can be taken from other standards on broadband traffic, particularly ITU-T Recommendation Q.823 on traffic management.

The information element is used in replies to requests for bearer allocation, modification, adding and dropping branches, and deallocation. The absence of the information element in the messages that could contain it indicates that congestion is not present.

This coding of this information element comes a close second to that of the repeat indicator for the Hamlet award for long-windedness. It loses to the repeat indicator because it actually contains useful information.

16.4.2.7 The Reject Cause

The reject cause information element has a single octet in its payload and this contains the reject cause type field. The values of the reject cause type field are used to indicate a large number of reasons for the rejection of a bearer connection control request from the SN.

This information element must be present in every rejection message sent by the AN. This is the only message specific information element sent in rejection messages, except when it is necessary to inform the SN of congestion in the rejection of allocations, modifications, or of the addition or dropping of branches.

16.4.2.8 The Branch Identifier

The branch identifier has two octets in its payload and these form the branch identifier value field. The values of this field uniquely identify the individual branches of a particular point-to-multipoint connection.

16.4.2.9 The Branch Identifier List

The payload of the branch identifier list contains one or more sets of two octets. Each set of two octets forms a branch identifier value field as used in the branch identifier information element.

The branch identifier list identifies one or more branches of a particular bearer connection. This information element is only used for requests to drop one or more branches of a connection.

16.4.2.10 B-BCC Information Element Anomalies

There is no need to have both a connection reference number and a connection reference number list because the connection reference number list can identify a single connection as efficiently as a connection reference number. If the reason for two difference information elements is to avoid the need to specify that the list has only one entry in certain cases, then the same logic should apply to other information elements that are flexible and these should also be split to avoid the need for to specify how they are used.

The same reasoning applies to the branch identifier. This is also unnecessary since the branch identifier list can identify a single branch as efficiently as the branch identifier.

16.4.3   B-BCC Protocol Anomalies

The inability of certain responses to the bearer channel requests to carry congestion information is difficult to understand since it cannot be known in advance that there is no need for congestion information. It would also be better if there was a specific congestion message which the AN could send spontaneously so that the SN does not have to wait until a response is sent to certain of its requests before it is notified of congestion. Because of both of these limitations, congestion notifications between the operations systems are more important.

It is strange that the allocation acceptance message (ALLOC_ACC) must contain the branch identifier if the request was for a point-to-multipoint connection because this ought to be the same as in the request as this is not a negotiable parameter but a reference. It would be better if the acceptance message did not include the branch identifier since it is already identified for the transaction and returning it only creates an opportunity for error.

The method of specifying the connection points at the service and user ports in an allocation request (ALLOC) is also quaint because a specific VC is specified at the service port, whereas at the user port it is also possible to specify just the port or the VP. The further negotiation here is even quainter since the request may specify optional VPs, but not optional VCs within them.

It is also curious that in the allocation completion message (ALLOC_COMP) the SN must specify the VC at the user port for the connection if this has not been previously agreed, but not the VC at the service port if that has not been previously agreed.

The allocation negotiation as a whole seems to be flawed because the AN is able to abort the process at either stage with a rejection message while the SN can only refuse to talk to the AN or issue a deallocation request (DEALLOC).

There is no need for the SN to identify the VC in the user port in the update branch (UPDATE_BRANCH) request because this can only be the VC identified in the acceptance of the add branch request, either explicitly or by tacit agreement with the request.

A similar approach to that of the BCC protocol for the narrowband V5 interface could have been used for unknown resources. It is curious that instead of this, no indication of unknown resources is passed back to the exchange, and this can be misleading since a positive acknowledgement without any indication of an inability to comply can be returned as the response to a request.

The absence of an auditing process in the B-BCC protocol makes the synchronization of connection between the AN and the SN more difficult, especially after the B-BCC protocol has been restarted. Effective operation of the protocol has to rely on co-operation between the management systems of the AN and the SN.

16.5 Summary

The difference between the VB5.1 and VB5.2 interfaces is manifest in the protocols that they use. Both type of interface use the Real-Time Management Coordination (RTMC) protocol, but only VB5.2 interfaces use the Broadband Bearer Channel Connection (B-BCC) protocol.

Both protocols use a common format, starting with a protocol discriminator. This is followed by the transaction octets, the message header octets, and then the message specific information elements. The message specific information elements also use a common format. Protocol errors are handled by an error message that has only formal differences when used in the different protocols.

The RTMC protocol was devised because it became clear that there were requirements for real-time coordination that could not be handled by the OAM flows described in ITU-T Recommendation I.610. There are three sets of messages in the RTMC protocol. One set deals with the blocking and clearing of connections. The second set deals with consistency checking of connections. The third set is for resetting the interface and checking the interface's identity.

The B-BCC protocol extends the functionality of the interface by adding dynamic connection control. The protocol enables bearer connections to be allocated, modified and deallocated. Branches to connections can also be added and dropped. B-BCC messages are normally organized into a request and associated acceptance and rejection messaged. The exception to this rule is the reporting of faults in the AN. The B-BCC protocol also includes housekeeping messages for resetting and presynchronization.

A number of anomalies and curiosities that have no technical justification are present in the VB5 protocols.

References

[1] *V-Interfaces at the Service Node (SN) - VB5.1 Reference Point Specification,* ITU-T Recommendation G.967.1, 1998.

[2] *V-Interfaces at the Service Node (SN) - VB5.2 Reference Point Specification,* ITU-T Recommendation G.967.2, 1998.

[3]   *V-Interfaces at the digital local exchange (LE) - V5.1- interface (based on 2048 kbit/s) for the support of access network (AN)*, ITU-T Recommendation G.964, 1994.

[4]   *V-Interfaces at the digital local exchange (LE) - V5.2- interface (based on 2,048 Kbps) for the support of access network (AN)*, ITU-T Recommendation G.965, 1995.

[5]   *B-ISDN Operations and Maintenance: Principles and Functions*, ITU-T Recommendation I.610, 1995.

17

VB5 Management

All is flux, nothing is stationary.
— Heraclitus

The ancient Greeks loved a good tragedy, and for many the sin of choice that led to tragedy was the sin of hubris. Hubris is an old pagan sin that is not very fashionable today, so it is good to discover that it still has its adherents. The difficulty with hubris today is the problem of finding the Gods who will visit nemesis on those presumptive mortals who take excessive pride in their own achievements.

The VB5 solution to this problem was to use a natural phenomenon, rather than ancient pagan Gods. This natural phenomenon is the one observed by Heraclitus and it is well suited for the punishment of hubris, both because of its classical origin and because the denial that change is eternal is itself a mark of hubris. In the VB5 case the agent of nemesis is the Internet Protocol, and it manifests itself in the management of VB5 interfaces.

17.1 Background

VB5 interfaces (see Chapters 15 and 16) have an architectural model and protocols, and they need to be managed [1, 2, 3]. This management [4, 5, 6] involves configuration, testing and fault handling of the Access Network (AN) and the Service Node (SN) over their Q3 interfaces (see Figure 17.1). The management also involves the co-ordination of the Operations System (OS) of the AN and of the SN over an X-VB5 interface.

There may be a large number of Virtual Path Connections (VPCs) between the AN and the SN. These must be grouped into Logical Service Ports (LSPs) that correspond to the VB5 interfaces (see Figure 17.2). Likewise, the VPCs between end-users and the AN must be grouped into Logical User Ports (LUPs), which are in turn mapped onto User Network Interface (UNI) accesses in the SN.

One of the VPCs between the AN and the SN must be configured to carry the VB5 protocols. One of the Virtual Channel Connections (VCCs)

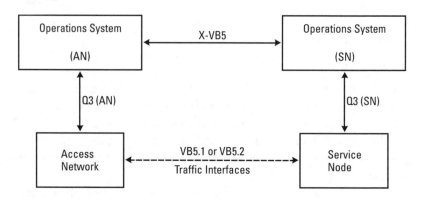

Figure 17.1 VB5 traffic and management interfaces.

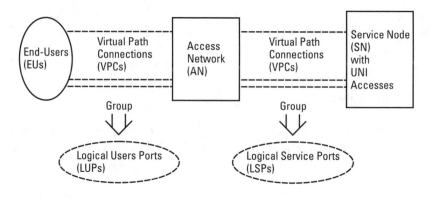

Figure 17.2 VPCs and logical ports.

within this VPC carries the Real-Time Management Co-ordination (RTMC) protocol (see Chapter 16). For a VB5.2 interface this VPC must also contain a second VCC that carries the Broadband Bearer Channel Connection (B-BCC) protocol.

The configurations in the AN and the SN must match and be co-ordinated even if the AN and the SN are managed by different network operators. The OSs on each side must be aware of the status of the VB5 resources on the other side, especially when the VB5 interface is not operational.

17.2   VPs, Logical User Ports, and Logical Service Ports

The AN and the SN associate VPCs with a VB5 interface in different ways because the AN has no knowledge of how the end-user signaling is configured. In both cases the LSP directly identifies and labels the end-points of VPCs that are not directly related to end-users. Likewise, in both cases the relationship between end-users and their LSPs involves two steps.

17.2.1   VP Level Configuration in the Service Node

The AN and the SN differ in way that the two-step relationships between end-users and their LSPs are modeled. In the SN there are UNI accesses (see Chapter 6) that are similar to those for directly connected end-users (see Figure 17.3) and these identify and label the end-points (the Trail Termination Points or TTPs), of VPCs in the same way. For VB5, they also reflect the status of the VPCs in the AN as indicated by the RTMC protocol.

A VB5 UNI access also indicates its LUP number and identifies its LSP. In addition, it identifies the end-points of the VCCs that carry user signaling, as for a direct access, but it may also identify intermediate points of VPCs. Identification of these intermediate points is included to support cross-connection of VPs in the SN.

VPCs from end-users terminate in the AN when the AN performs VC cross-connection. This is reflected in the SN by a VPC LUP object that is contained directly in the UNI access because the VPC from the end-user does continue on to the SN. In this case, the VPC LUP object reflects the state of the remote end user VPC as indicated by the RTMC protocol.

If the VC cross-connection in the AN is controlled by the B-BCC protocol then the VPC LUP object in the SN also holds traffic descriptors and quality of service parameters. These are needed because the SN must

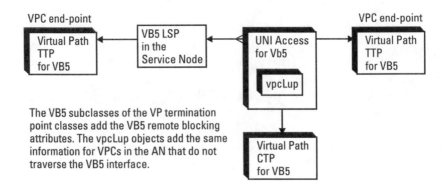

Figure 17.3 VP level configuration in the service node.

know whether the VPC between the AN and the end-user can handle a particular call. The VPC LUP also indicates whether the remote VPC can contain VCCs that are semi-permanently configured in addition to VCCs that are controlled by the B-BCC protocol. In this case the VPC LUP contains Virtual Channel Link (VCL) information for the LUP (see Figure 17.4).

The VCLs at the LUP indicate if the semi-permanent connection in the AN is controlled by the B-BCC. In this case then the SN will attempt to re-establish the connection using the B-BCC after a failure in the AN. The VCLs also indicate the Virtual Connection Identifier (VCI) at the LUP and identify the corresponding intermediate point of the VCC in the SN.

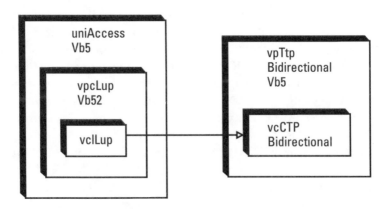

Figure 17.4 Modeling of mixed remote VPCs for VB5.2.

17.2.2 VP Level Configuration in the Access Network

In the AN there are no UNI accesses because the AN does not know about user signaling. Instead the AN contains LUP objects (see Figure 17.5).

The roles of the intermediate points and end-points of VPCs that are associated with end-users in the AN are reversed from those in the SN. This is because it is natural to have intermediate points of VPCs from end-users in the AN and for the end-points of these connections to be in the SN. The modeling in the AN is not enhanced because the RTMC protocol only carries status information to the SN.

The modeling of the Transmission Convergence (TC) in the AN needs to be modified when a VB5 interface is present to let the OS of the AN initiate the clearing of connections by triggering the sending of an AWAIT_CLEAR message to the SN. This is modeled by an extension that is contained in the normal TC adaptor (see Figure 17.6).

17.2.3 Configuration of VB5 Protocols

The LSP always contains an RTMC Communications Path because VB5.1 and VB5.2 interfaces both have an RTMC protocol (see Figure 17.7). The

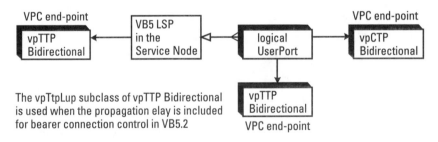

Figure 17.5 VP level configuration in the access network.

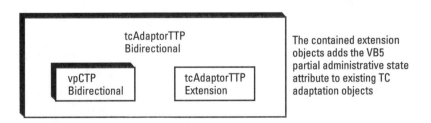

Figure 17.6 Extending the transmission convergence in the AN.

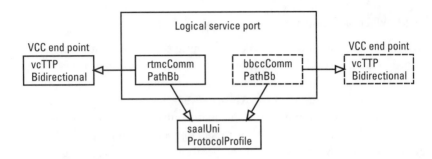

Figure 17.7 VB5 Protocol configuration.

RTMC Communications Path identifies the end-point of the VCC that car-ries the RTMC and identifies the profile that is used for the Signaling ATM Adaptation Layer (SAAL). If the VB5 interface is a VB5.2 interface then the LSP also contains a B-BCC Communications Path that identifies the VCC end-point and the SAAL profile for the B-BCC.

17.3   The Relationship with VB5 Messages

Many VB5 messages are related to transactions on the management inter-faces. Sometimes VB5 messages are triggered by the activities of an OS. At other times the VB5 messages cause notifications to be sent to an OS.

17.3.1   Start-Up of an Interface

Although a start-up may be initiated autonomously by the AN or the SN following a failure of the SAAL, the focus here is on a start-up initiated by an OS. If the LSP is already restarting or is activated there is an immediate response to the OS and no additional action is taken. If the LSP is inactive then a new start-up is initiated. The first step is to boot up the SAAL. Once the SAAL is established the AN and the SN can exchange VB5 messages.

The next step is an RTMC start-up. This involves checking the iden-tity of the VB5 interface and performing a RTMC reset. If this is successful on a VB5.1 interface then the activation state of the interface changes to acti-vated and the OSs are notified.

For a VB5.2 interface the successful completion of the RTMC reset results in the operational state of the interface changing to enabled, but a B-BCC reset needs to occur before the interface becomes fully activated.

17.3.2 Checking the Interface

The LSPs in the AN and the SN support OS requests to check the identity of the VB5 interface. This triggers the VB5 REQ_LSPID message that requests the other side of the interface to indicate how it has labeled the interface. The response to the OS indicates if the labels on both sides match.

An LSP in the SN can support consistency checking of Virtual Path Connection Identifiers (VPCIs) using loopback cells. This applies to VPCs on a VB5 interface that terminate in the AN because the VB5 interface cannot guarantee the return of loopback cells from an end-user. Following a requested VPCI check, the SN will indicate the result to its OS, including the nature of any failure.

17.3.3 Resetting the VB5 Interface

The OSs can initiate an RTMC reset either directly or indirectly as part of a start-up. This causes the states of the LSP and the VPCs to be reset. On-demand connections may also be released, although the SN should attempt to minimize the interruption to service. The OSs on both sides of the interface are informed of the result of the reset.

The B-BCC reset is asymmetric and cannot be triggered by the OS of the AN. This is a reflection of the asymmetry of the B-BCC protocol itself, where the SN is the master and the AN is the slave. The OS of the SN receives a response if it has initiated a B-BCC reset, and it is notified if the reset was initiated autonomously by the SN. The OS of the AN can only be notified of the reset.

17.3.4 The State of Resources

The RTMC protocol has clearing messages that allow resources to be shut down without disturbing on-going calls. It also has messages that indicate if resources are blocked or unblocked.

The OS of the AN can request resources in the AN to be shut down. This can trigger the exchange of clearing messages between the AN and the SN. If this is successful then the resources will become administratively locked.

When resources in the AN become locked, a block message is sent to the SN, which notifies its OS. A block message can also be sent to the SN if there is a fault in the AN.

17.4 Broadband Access Coordination: X-VB5

The RTMC protocol is not able to coordinate management across a VB5 interface when the interface is not operational. This is less of a problem if the AN and the SN are controlled by the same OS. It is a problem if the AN and the SN belong to different network operators because in this case a new interface is needed.

The X-VB5 interface [6] carries information between the OS of the AN and the OS of the SN. Many of its transactions coordinate the addition and removal of resources. The remaining transactions are more specific.

17.4.1 Specific X-VB5 Transactions

The specific X-VB5 transactions coordinate the labeling of VB5 interfaces, allow their configuration to be verified, and support the notification and localization of faults (see Figure 17.8).

When a VB5 interface is being created the OSs need to agree on how it will be labeled. The X-VB5 interface allows the OSs to agree a unique identifier by concatenating the local labels by which the AN and SN refer to

Catagory	Name	Source	Equivalent CMIP Type
Labels and References	anLabelRequest	AN	Action
	snAndLabelRequest	SN	Action
Verification: Audit Requests	auditConnectionRequest	Either	Action
	auditVpciRequest	Either	Action
Verification: List Requests	listLupsRequest	Either	Action
	listProtocolsRequest	Either	Action
	listVb5InterfacesRequest	Either	Action
	listVpsRequest	Either	Action
	listVpsRequest	Either	Action
Fault Notification	resourceStatusIndication	Either	Notification
Fault Localization	addLoopRequest	SN	Action
	removeLoopRequest	SN	Action

Figure 17.8 Specific X-VB5 transactions.

each other and the local reference by which the AN refers to the VB5 interface.

The verification requests consist of audit requests and list requests. The audit requests allow either OS to request the other to return the complementary information to that given in the request. The connection audit asks for the other port identifier, VPI and, if appropriate, the VCI. The VPCI audit asks for the VPCI that corresponds to the port and ATM identifiers, or vice versa. The list request asks for the list of VB5 interfaces, or for the details of a specified VB5 interface.

The fault notification transaction allows either OS to inform the other that the status of a resource has changed. The fault localization transactions allow the OS of the SN to request the AN to add or remove an ATM loop.

17.4.2 Generalized X-VB5 Transactions

Either OS can request the addition or removal of a resource and either OS can notify the other that a resource has been added or removed (see Figure 17.9). Thus for each type of resource there are a total of four transactions, i.e. requests and notifications of addition and removal.

There are both notifications and requests because resources can be added and removed autonomously. When an OS receives a notification of an autonomous change, it may treat it as a request to perform an equivalent change. The OS that performs an autonomous change may also send a

Function	Item	Message Type
Add	Vb5Interface ProtocolVp Vb5Protocol	Request = Action
Remove	Lups Vps Vcs Connection	Indication = Notification

Figure 17.9 Generalized X-VB5 transactions, used by either OS.

corresponding request to the other OS. OSs may inform each other of the change of status of a resource before it is removed using the specific fault notification.

17.4.3 RPC Specification of X-VB5 Transaction Requirements

The requirements for the transactions of the X-VB5 interface are formally specified using Remote Procedure Calls (RPCs) [7] since this allows the information they carry to be clearly defined. RPC is an effective way of specifying requests because the request and the response can carry different information. It is also effective for notifications because the client and agent roles are specific to each call and the response confirms that the notification has been received.

If a CMIP implementation of the X-VB5 interface is desired, then the RPC definition of the required transactions has an easy mapping onto CMIP actions and notifications. Using RPC also helps to future-proof the X-VB5 interface by allowing implementation protocols other that CMIP to be used.

Since the RPC protocol is a TCP/IP application protocol, its use ensures that the X-VB5 interface is compatible with IP. The parameter definitions that are used in the transaction specifications are not optimized for an RPC implementation of X-VB5 because they have been structured for clarity instead of efficiency. This does not prevent their use, and they could be easily restructured along the lines of the CMIP implementation.

Using RPC to ensure that the X-VB5 interface is not made obsolete creates a threat to the conventional VB5 interface that is more insidious that obsolescence. The X-VB5 interface has to include many of the functions of the VB5 RTMC protocol since it must support management co-ordination when the VB5 interface is not operational. The X-VB5 interface must also co-ordinate the addition and removal of the permanent ATM connections that are not controlled by the VB5 B-BCC protocol.

This creates the possibility of using X-VB5 as a real-time IP signaling protocol with similar functions to the RTMC and B-BCC protocols. This would be fast and simple to implement and easy to extend. Such an effective X-VB5 interface could undermine the use of the RTMC and B-BCC protocols. Although this would be a blow to the conventional wisdom on VB5 signaling, it could be in its long-term interest because it would make VB5 compatible with IP.

17.5 Summary

The management of VB5 involves Q3 interfaces between the Access Network (AN) and its Operations System (OS) and between the Service Node (SN) and its OS. It also involves the X-VB5 interface between the OSs themselves. The Q3 transactions support the configuration, testing and fault reporting of the AN and the SN. The X-VB5 transactions support the co-ordination of these functions.

To configure a VB5 interface, an OS must group Virtual Path Connections (VPCs) between the AN and the SN together to form a Logical Service Port (LSP). VPCs that terminate in the AN are associated directly with the LSP. VPCs that go via the AN to the end user are grouped into Logical User Ports (LUPs). In the AN, LUPs are associated in turn with LSPs, but in the SN, User Network Interface (UNI) accesses are used instead of LUPs because the configuration in the SN includes user signaling.

An OS configures the VB5 protocols in the LSP by associating the end-point of a Virtual Channel Connection (VCC) with the profile for a Signaling ATM Adaptation Layer (SAAL). There is always an association for the Real-Time Management Co-ordination (RTMC) protocol. For VB5.2 interfaces there is a second association for the Broadband Bearer Channel Connection (B-BCC).

The Q3 transactions are related to transactions across the VB5 traffic interfaces. The OSs can initiate start-up of the VB5 interface, and this can lead to VB5 messages that check the interface and reset its resources and protocols. Q3 transactions also allow the OSs to take VB5 resources in and out of operation, and notify the OSs of faults and remote conditions.

The X-VB5 interface has both specific transactions and generalized transactions. The specific transactions are used for co-ordinated VB5 labeling, for auditing and verification, and for remote fault reporting when the VB5 interface is not operational. The generalized transactions support the co-ordinated addition or removal of the resources.

The X-VB5 requirements have been specified using the Remote Procedure Call (RPC) protocol so that the requirements are well defined and can be mapped onto implementation protocols. RPC is an IP application protocol, and this raises the question of whether the RTMC and B-BCC protocols are really necessary since the X-VB5 interface could easily provide similar functionality using IP based signaling since RPC could also be used as the X-VB5 implementation protocol.

References

[1]   *V Interfaces at the Digital Service Node - VB5.1 Reference Point,* ETSI Standard EN 301 005-1, 1997.

[2]   *V-Interfaces at the Service Node (SN) - VB5.1 Reference Point Specification,* ITU-T Recommendation G.967.1, 1998.

[3]   *V-Interfaces at the Service Node (SN) - VB5.2 Reference Point Specification,* ITU-T Recommendation G.967.2, 1998.

[4]   *VB5.1 Management,* ITU-T Recommendation Q.832.1, 1998.

[5]   *VB5.2 Management,* ITU-T Recommendation Q.832.2, 1999.

[6]   *Broadband Access Co-ordination,* ITU-T Recommendation Q.832.3, February 2000.

[7]   Srinivasan, R., *RPC: Remote Procedure Call Protocol Specification Version 2,* RFC 1831, (http://www.ietf.org/rfc/rfc1831.txt), August 1995.

18

Optical Access

Into the Valley of Death rode the 600.
—Alfred Lord Tennyson, *Charge of the Light Brigade*

According to the Bible, the creation of the Universe began when God said, "Let there be Light." Recent cosmological theories are less positive. In particular, the idea that 99% of the Universe consists of dark matter that was created immediately after the Big Bang suggests that God may in fact have said, "Let there be Dark."

The development of broadband access for end-users has followed a similar pattern. The conventional wisdom of the 1980s was that this would begin with light, flowing within optical fibers. By the 1990s it was accepted that most broadband accesses would use the darker art of ADSL.

However, the unwavering faith of ancient patriarchs in optical technology will eventually be rewarded. Like the 600, they also seem to have fallen in the valley of death, but they will rise again and inherit the Earth on that day of reckoning when the bandwidth runs out.

18.1 Background

The current wisdom is that broadband access for end-users will be initially delivered using Asymmetrical Digital Subscriber Line (ADSL) technology.

ADSL is effective because it operates over the same lines that deliver ordinary telephony service and can be added on top of the service. ADSL can also operate over the full length of most lines.

ADSL is a better prospect than cable modems because cable television operates over a shared medium, whereas each telephone line is dedicated to an end-user. To overcome this limitation, strategic investment is needed to change the cable television architecture to hybrid fiber-coax since this reduces the number of end-users that share the coaxial bandwidth. Cable modems also suffer because cable television is not as universal a service as telephony.

The effectiveness of ADSL has created its own limitations. In particular, its data rate has been constrained to maximize its deployment. For higher data rates a better transmission medium is needed for at least part of the access link, and optical fiber is the obvious choice.

ADSL has delayed the introduction of optical technology because it supports broadband access over the existing telephony infrastructure. In the long term this delay may be beneficial because it has allowed applications to become more attractive and standards to become more mature. Credit for encouraging the development of optical technology goes to the FSAN (Full Services Access Network) collaboration, which drafted the specifications that have been subsequently downstreamed into more conventional standards organizations [1, 2].

18.2   ATM PON Architecture

The architecture of ATM Passive Optical Networks (PONs) supports both pure optical accesses and hybrid optical accesses with final links over metallic wires. The interface to the core network is at an Optical Line Termination (OLT) (see Figure 18.1) and the passive optical medium that an OLT connects to is called the Optical Distribution Network (ODN). A single OLT can connect to a number of ODNs.

Typically the ODN has several remote ends, known as Optical Network Units (ONUs), since these can be supported by a single laser and receiver at the OLT. In the case of a pure optical access, the ONUs are referred to as Optical Network Terminations (ONTs) because they include the Network Termination (NT) functionality. In the case of a hybrid optical access, the NTs are connected to the ONUs by metallic wires.

A pure optical architecture is often called FTTH (Fiber To The Home) because business applications were originally thought of as hybrid accesses.

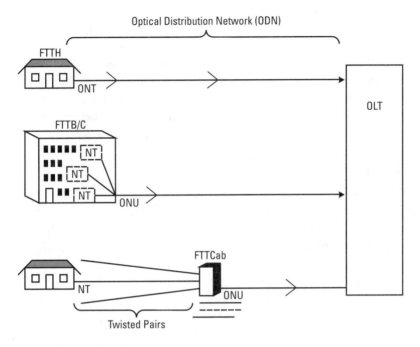

Figure 18.1 ATM PON architecture.

Hybrid business architectures are now included within FTTB/C (Fiber To The Building/Curb). FTTB/C has short wire drops that are capable of operating at very high speeds. If the final wire drop is lengthy then the hybrid architecture is called FTTCab (Fiber To The Cabinet).

18.2.1 ATM PON Transmission

The ITU-T recommendation on ATM Passive Optical Networks (PONs) [1] covers the physical medium and optical devices, the data rates and frame formats, and the ranging and bandwidth allocation for point-to-multipoint operation.

ATM PON transmission uses SONET/SDH rates, but it has both symmetric and asymmetric transmission modes. It is better suited for the links into an ATM core network than conventional SONET/SDH because it is less wasteful of bandwidth when the traffic is asymmetric and because it has a simple frame format based on ATM cells.

The upstream rate is always 155.52 Mbps. Like ASDL transmission, the asymmetric mode is faster downstream than upstream, but unlike ADSL the downstream rate (622.08 MBit/s) is fixed at four times the upstream rate. The upstream frame format is different from the downstream format to allow traffic from the different ONUs to merge (see Figure 18.2).

Although the data rates are much less than the limit for optical fiber, this is not a major problem because there could be parallel transmission on different optical wavelengths. Higher rates could also be defined either by adding more cells per frame or by transmitting the frames at a higher rate.

18.2.1.1   Frame Formats

The different frame formats in the different directions give less payload capacity in the upstream direction than in the downstream. Frames contain cell slots, but normal ATM cells are transmitted continuously downstream whereas the upstream slots contain three additional overhead octets. These octets are used for guard time, the preamble, and the delimiter.

The guard time ensures that cells from different ONUs do not overlap when they merge. The preamble allows the recovery of the timing and the amplitude of the bits. The delimiter allows the recovery of the start of the conventional cell boundary.

In symmetrical transmission there are the same number of octets in each frame. In the downstream direction there are 56 slots of 53 octets each. If the downstream traffic is at 622.08 Mbps then the content of the 155.52 frame is repeated four times. In the upstream direction there are 53 slots of 56 octets because of the three octets overhead.

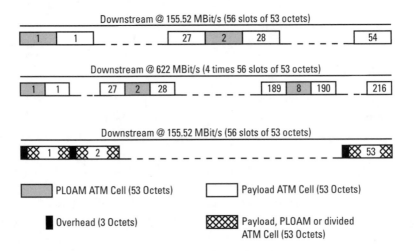

Figure 18.2 PON frame formats.

Downstream traffic is broadcast to all ONUs. The OLT controls the allocation of cell slots for upstream transmission through the downstream Physical Layer Operations, Administration, and Maintenance (PLOAM) cells. In the downstream direction every 28th cell is a PLOAM cell. The OLT also controls which upstream cells are PLOAM cells, which are payload cells, and which cells are divided. The difference between the allocation of cell slots in the two directions gives flexibility in the allocation of PLOAM cells to the ONUs while guaranteeing an unambiguous flow of PLOAM cells to the ONUs.

18.2.1.2 Divided Cell Slots

Often the OLT allocates a number of cell slots to an ONU. This is not always desirable because it means that the smallest unit of bandwidth that can be allocated to an ONU is over 2.5 Mbps.

To avoid this the OLT can divide an upstream cell slot into mini slots that are allocated to different ONUs. Like the upstream cell slots, each mini slot begins with three overhead octets. The payload of a mini slot can range from one to 53 octets

18.2.1.3 Security

To increase security, each ONU can send a key upstream to its OLT. Upstream transmission is more secure than downstream transmission because it is only received by the OLT. The OLT uses the key to scramble the data sent to the ONU. The ONUs also send the OLT their passwords so that the OLT can check that a malicious ONU is not masquerading as different ONU.

18.2.2 VDSL Transmission on Hybrid Architectures

Very-high-speed Digital Subscriber Line (VDSL) transmission can operate at higher rates than ADSL. Downstream rates up to 52 Mbit/s are possible [3], and both Discrete MultiTone (DMT) and Carrierless Amplitude and Phase (CAP) techniques have been considered. The higher VDSL rates reduce the transmission distance, especially for symmetric transmission, but the reach is sufficient to allow VDSL to be used on the metallic links in a hybrid architecture (see Figure 18.1).

VDSL is made possible by advances in Digital Signal Processing (DSP) that allow sophisticated modulation, equalization and echo-cancellation techniques to be implemented. Its main limitation is cross talk. Near-End Cross Talk (NEXT), which is produced by transmission at the same end of the line, has a greater effect than Far-End Cross Talk from a transmission at the remote end. This is because FEXT is attenuated more by transmission. Asymmetrical transmission makes greater distances possible because much of the high frequency NEXT can be filtered out.

VDSl is also limited by impulsive noise, such as the transients produced by crude telephony signaling, and interference, such as that from radio sources. DSP also introduces effective noise because of the limited precision of the arithmetic and of the analogue to digital conversion.

VDSL transmission should not interfere with narrowband transmission on the same line so that it can be deployed on lines that carry existing traffic. VDSL should also be compatible with other DSL technologies on other pairs in the same cable.

The standardization of VDSL has been slow because of the greater incentive to develop ADSL, which can operate on most lines, and optical transmission, which is needed regardless of VDSL. Many operators have also expressed reservations about hybrid architectures because they require equipment to be deployed in hostile environments.

18.3   PLOAM Cells on ATM PONs

PLOAM cells perform the physical layer management of the ATM PON. The OAM flow that they carry is an F3 flow, so it sits immediately below the F4 and F5 ATM layer flows. PLOAM cells can be identified by the value of their Virtual Path Identifier (VPI). Unlike F4 and F5 flows, there is no differentiation here between end-to-end and segment flows so only one VPI value is needed.

The VPI and the Virtual Channel Identifier (VCI) of these cells are both zero (see Figure 18.3). The most significant bit of the PT field is one to indicate that these are a management overhead, making the PT field the same as that for an F5 segment flow. The CLP (Cell Loss Priority) bit is also one, but this has no real meaning because cells are simply discarded at the ends of the optical transmission.

The upstream and downstream PLOAM cells have different payloads (see Figure 18.4). Some of the fields are common to both upstream and downstream cells, while others are specific to the directions.

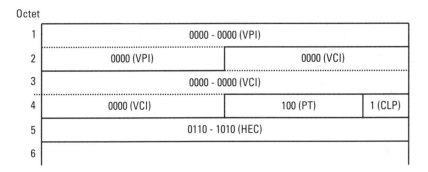

Figure 18.3 Header for PLOAM cells.

18.3.1 Common PLOAM Fields

The common PLOAM fields are the IDENT, Message, and BIP fields (see Figure 18.4). A Cyclic Redundancy Check (CRC) follows the Message field that it validates.

18.3.1.1 IDENT Field

ONUs need to know the location of the downstream frame boundaries tod etermine the numbering of the cells. The cell header allows the PLOAM cells to be detected, and the least significant bit of the IDENT field [1]

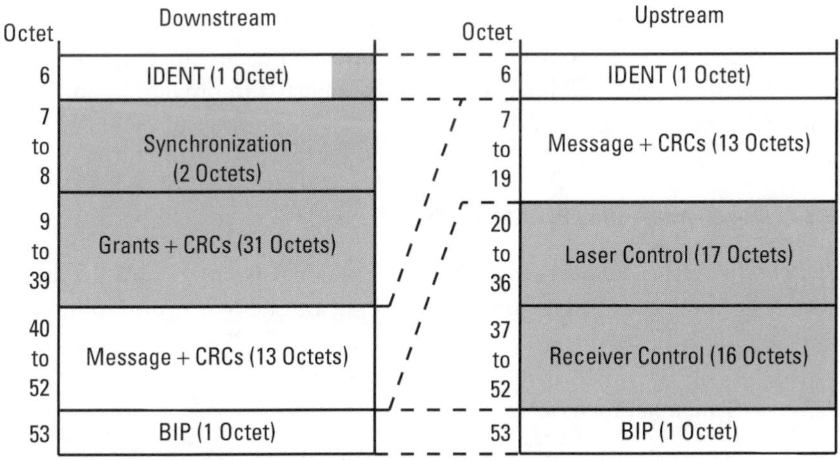

Figure 18.4 PLOAM payload fields.

carries the frame alignments signal in the downstream direction (see Figure 18.4). This bit is set to one in the first PLOAM cell of a downstream frame and is set to zero in all other frames.

The ranging process determines the frame alignment in the upstream direction.

18.3.1.2   Message Field

The OLT must be able to talk to each of the ONUs and each of the ONUs must be able to reply to it. The message field allows this, and in particular it supports ranging and event notification.

In the Message field, the identifier and message type precede the message contents. The identifier is the number that is assigned to an ONU during the ranging phase. The OLT uses this number to address messages to different ONUs and the ONUs use this number to identify themselves when they respond. The OLT can also use a special address to broadcast messages to all ONUs.

The OLT does not use the Message field directly to grant upstream cell slots to ONUs. Instead it uses the Message field to instruct the ONU to examine the Grants field.

18.3.1.3   BIP Field

The PON transmission needs to have some way of monitoring its performance, and the BIP (Bit Interleaved Parity) field in the PLOAM cells provides this. Each bit of this octet is a parity bit for the corresponding bits in the payload octets between it and the preceding BIP octet. Although this approach does not detect all errors, it is sensitive to bursts of errors since neighboring errors affect different BIP bits, and it can be adjusted to provide a good estimate of the BER (bit error rate).

18.3.2   Downstream-Only PLOAM Fields

The two fields that follow the IDENT field in downstream PLOAM cells are specific to these cells (see Figure 18.4). These are the Synchronization field and the Grants field.

18.3.2.1   Synchronization Field

For some applications it is desirable to convey an accurate timing reference from the OLT to the ONUs. The two octets of the Synchronization field provide a means of doing this.

The method used is similar to the Synchronous Residual Time Stamp (SRTS) method of AAL1. In both cases there is a counter and its least significant bits are carried to the other end. In this case a 1 kHz reference resets the counter. If the downstream transmission rate is 155.52 MBit/s then the counter is clocked each time a cell is transmitted, but if the rate is 622.08 MBit/s then the counter is clocked once every four transmitted cells.

18.3.2.2   Grants Field

The upstream cell slots and the mini slots of divided cells must be allocated to ONUs. Cell slots must also be assigned to traffic and to PLOAM cells.

The first two downstream PLOAM cells in each frame carry the grants for all of the upstream cell slots. Each of these PLOAM cells carries grants for 27 slots. After each group of seven grants, and after the last group of six grants, there is a CRC field that validates the preceding grants.

Values are used to indicate cell slots that are unassigned, idle or for use in ranging. The last grant in the second PLOAM cell has the idle value since there is no 54th cell in the upstream direction. If the downstream transmission is at 622.08 MBit/s then the grants in the remaining six PLOAM cells are also marked as idle since there are no corresponding upstream cells. Messages carried in the Message field of PLOAM cells determine how the grants for each cell slot are used.

18.3.3   Upstream-Only PLOAM Fields

The two fields that are specific to upstream cells are the Laser Control and Receiver Control fields (see Figure 18.4).

18.3.3.1   Laser Control Field

The lasers in the ONUs of different implementations will have different characteristics and the mean optical power level may vary from the required level. The laser control field is a pattern that is sent by the ONUs to maintain the mean optical power level and to help the OLT determine when the ONUs should be allowed to transmit. The form of this field is not specified since it depends on the implementation.

18.3.3.2   Receiver Control Field

The strength of the optical signal that the OLT receives differs from ONU to ONU because the signal paths are different. This means that optimum threshold for differentiating between received ones and zeros is different for each ONU.

To help the OLT set these thresholds, each ONU transmits a defined pattern in the Receiver Control field. There is a default pattern or the OLT can also use the Message field to define a different pattern.

18.4   OLT/ONU Coordination

Initial communication between ONUs and OLTs must be based on serial numbers since these uniquely identify the ONUs. An ONU can inform the OLT of its serial number when it is in the ranging mode. The OLT needs to know the serial number of an ONU to initially address it. It can then send a message to the ONU to grant it access to upstream cell slots for ranging.

The OLT can also assign a PON identifier to the ONU. The OLT can use this identifier to deactivate transmission from an ONU, or it can use the serial number of the ONU to disable the ONU from participating in the ranging process.

There is a wide range of co-ordination functions that can be used during normal operation. These include error monitoring, event notification, continuity checking, and security. There is also scope for vendor specific extensions.

18.5   Summary

A new optical fiber infrastructure and ATM PON (Passive Optical Network) technology allows higher transmission rates than can be achieved with ADSL (Asymmetrical Digital Subscriber Line). The introduction of ATM PONs has been delayed because ADSL supports broadband access over the same line that simultaneously delivers telephony service, but this delay may be beneficial because it has allowed applications to develop and standards to mature. The FSAN (Full Services Access Network) collaboration has been instrumental in encouraging the development of these standards.

ATM PONs support both pure optical access, sometimes called Fiber To The Home (FTTH), and hybrid access with metallic transmission on the final link. Hybrid access is called FTTB/C (Fiber To The Building / Curb) if the final metallic link is short, or FTTCab (Fiber To The Cabinet) if the link is long. An Optical Line Termination (OLT) converts the transmission link from the core broadband network onto transmission over a number of passive Optical Distribution Networks (ODNs). Each ODN can support a number of remote ends, called Optical Network Units (ONUs). ONUs are called Optical Network Terminations (ONTs) in a pure optical architecture.

The final link in a hybrid architecture may use Very-high-speed Digital Subscriber Line (VDSL) transmission. This supports data rates up to 52 MBit/s, limited mainly by crosstalk from other transmissions in the same cable. VDSL standardization has lagged behind that of ADSL and ATM PONs because it cannot be used independently except on short lines. Some operators also have reservations about the use of hybrid architectures.

Upstream ATM PON transmission operates at 155.52 Mbps, but the downstream rate may be either 155.52 Mbps or 622.08 Mbps. Although these frame rates are compatible with SONET/SDH, the frame formats are simpler, cell based, and differ in each direction. There is always a smaller payload in the upstream direction even since capacity is needed to allow transmission from different ONUs to merge.

PLOAM (Physical Layer Operations, Administration, and Maintenance) cells that can be identified by their headers perform the physical layer management of ATM PONs. There are fixed PLOAM cell slots is the downstream transmission and these contain fields that grant upstream cell slots to ONUs. PLOAM cells indicate the downstream frame boundary and carry parity bits for performance monitoring. Downstream cells can also carry a timing reference signal. Upstream cells contain fields to allow the OLT to decide when ONUs should transmit and what thresholds to use to detect bits.

Coordination between ONUs and their OLT is needed for ranging and for the allocation of cell slots for upstream transmission. During normal operation the communication between ONUs and their OLT supports a wide range of functions, and this can be further increased by vendor specific extensions.

References

[1]   *High Speed Optical Access Systems based on Passive Optical Network (PON) Techniques,* ITU-T Recommendation G.983, 1998.

[2]   *ATM PON Management,* ITU-T Recommendations Q.834.1, Q.834.2, Q.834.3, approval scheduled 2001.

[3]   Chen, W. Y., "The Development and Standardization of Asymmetrical Digital Line," *IEEE Communications Magazine,* Vol. 37, No 5, May 1999, pp. 68–72.

19

ATM Enhancements

I have to admit it's getting better.
—The Beatles

In 1979 the Nobel Prize in Physics was won for work that showed how electromagnetism and the weak nuclear force are different aspects of the same phenomenon. Although light is a fundamental to electromagnetism, it was not necessary for Edison to wait for this breakthrough before he could invent the light bulb. Sometimes it seems that if this had been the story of ATM we would still be awaiting the invention of the light bulb, and the effort required to change one would be a matter for future study.

It is getting better. This chapter describes some of the ways that ATM has been recently improved. But it is not enough. There is still a need to boldly go where no one has gone before.

19.1 Background

The Operations, Administration and Maintenance (OAM) flows that were originally defined for the ATM layers have a number of deficiencies and this led to their modification [1]. The greatest innovation was the introduction of Automatic Protection Switching (APS) [2]. APS is particularly useful because it allows ATM traffic to be protected when it is carried over a simple cell based physical layer.

An alternative approach to increasing resilience is the use of call control signaling to establish soft permanent connections and to re-establish these in the event of a failure. This is assisted by the introduction of call control signaling for Virtual Path (VP) connections since this allows soft connections to also be established at the VP layer.

Work has also been carried out on multipoint connections and on the characterization of ATM connections. Unfortunately, these are areas where there the potential of agreement on a uniform approach between ITU-T and the ATM Forum seems small.

19.2   Enhanced OAM Flows

OAM flows were defined in the original version of ITU-T recommendation I.610 and were enhanced in the 1999 version [1]. In this version, the format of location identifiers were specified, an activation / deactivation function for forward performance monitoring was introduced, and cells for ATM Automatic Protection Switching (APS) were defined. The modified OAM cell types and function are summarized in Figure 19.1. Unfortunately, this is also a catalogue of missed opportunities.

Many changes are modifications to the function specific fields in the original OAM cells (see Chapter 6 on ATM OAM flows). The 16-octet location field in AIS, RDI, and loopback cells was clarified (see Figure 19.2). Unfortunately, the opportunity to define specific defects in the defect type field in AIS and RDI cells and to define the format of the correlation tag field, which matches transmitted and returned loopback cells, was lost.

Two fields were added to Backward Reporting cells. One contains the sequence number of the Forward Monitoring cell that the report is for. The other contains the current value of the counter of received cell blocks with severe errors. Unfortunately the opportunity to specify the format of the timestamp field in monitoring and reporting cells was lost. Although this omission is not critical, it leaves the field meaningless.

The issue of erroneous configuration of ATM circuits was not resolved. Although loopback cells can be used to test configuration, they do not provide positive in-service confirmation because loopback testing is not normally performed continuously. Performance Monitoring cells can provide positive in-service confirmation, but they require significant processing power and do not detect erroneous delivery of the cell stream to a third party.

OAM Cell Type	Cell Function	Description
0001 - Fault Management	0000 - AIS	Ongoing alarm indication signal
	0001 - RDI	Returned defect indication
	0100 - Continuity Check	Maintains minimum cell flow
	1000 - Loopback	Non-intrusive loopback cell
0010 - Performance Management	0000 - Forward Monitoring	Ongoing bit interleaved parity
	0001 - Backward Reporting	Returned performance report
0101 - APS Co-ordination Protocol	0000 - Group Production	Protection of groups of connections
	0001 - Individual Protection	Protection of individual connections
1000 - Activation/ Deactivation	0000 - Forward and Backward Performance Monitoring	Activates and deactivates both monitoring and reporting
	0001 - Continuity Check	Activates and deactivates continuity checking
	002 - Forward Performance Monitoring	Activates and deactivates performance monitoring

Note 1: The shaded entries indicate modifications.

Note 2: Although the Continuity Check cell has not yet been modified, an appendix to I.630 (1999) contains an annex on a proposed future modification.

Figure 19.1 Modified OAM cell types and functions.

Identifier Type (Octet 1)	Location Coding (Octets 2 to 16)
0000 - 0000	All zeros = 0000 - 0000.
0000 - 0001	Country Code + Network ID + Operator Specific Coding Octets 2 to 5: BCD: E.164 Country Code + Network ID Octets 6 to 16: Binary: Operator Specific Coding
0000 - 0010	Country Code + Network ID Octets 2 to 5: BCD: E.164 Country Code + Network ID Octets 6 to 16: Coded as 6A (Hex) = 0110 - 1010
0000 - 0011	Partial NSAP Coding (not fully defined)
0110 - 1010	All 6A (Hex) = 0110 - 1010
1111 - 1111	All ones = 1111 - 1111

Note: The different coding types cater for backwards compatibility and for ITU-T, ATM Forum and operator specific variations.

BCD = Binary Coded Decimal
Hex = Hexadecimal
NSAP = Network Service Access Point

Figure 19.2 Location identifier field.

A simple way to identify erroneous configuration is to use Continuity Check cells with unique source identifiers, and this could have used the newly clarified location identifier field. If these new Continuity Check cells were discarded when they were not used then they could be sent as a matter of course on all ATM connections. Received cells could be processed where and when required, creating a powerful tool to give in-service confirmation that connections exist and are correctly configured. Unfortunately this was yet another missed opportunity.

19.3   ATM (Automatic Protection Switching) APS

Protection switching at the ATM layer was defined to improve the reliability of ATM connections when the reliability of the physical transmission layer is not adequate. ATM protection switching may cause cells to arrive in the wrong order, and there may be loss or duplication of cells.

Since a physical failure affects a group of connections, ATM protection switching includes both the protection of groups and of individual connections. Protection of individual ATM connections can be used in conjunction with the protection of connection groups. ATM protection switching can also be used in conjunction with physical layer switching, in which case it should be delayed so that the physical protection can operate first.

ATM protection switching can operate at both ATM layers. If the protection is for only part of the range of the OAM flow then the APS OAM cells that carry the protection protocol must be monitored non-intrusively. The protected part of the connection or connection group that is used under normal conditions is known as the working entity. The backup that is used when the working entity fails is known as the protection entity.

19.3.1   Types of Protection Switching

ITU-T Recommendation I.630 [2] specifies different types of protection switching, namely bidirectional 1+1 and 1:1 protection, and unidirectional 1+1 protection. Unidirectional protection switching means that only the affected direction of transmission is switched. Bidirectional protection switching means that both directions of transmission are switched even if one of the directions is not affected.

In 1+1 protection, the transmission path is duplicated and the destination selects the better path. No protection protocol is needed if 1+1 protection is unidirectional because no communication with the source is required.

1:1 protection is a special case of M:N protection where M backups are provided for N working paths. Here both ends must agree to switch traffic to a backup when a protected path fails. In ATM 1:1 protection, the backup path may carry extra traffic that is not protected and is pre-empted when a protection switch occurs.

Protection switching can occur because of a detected failure, a manual intervention, or instructions from an Operations System. The switching can be revertive or nonrevertive. If it is revertive then transmission reverts to the original path once a fault is cleared. Nonrevertive switching is sometimes preferred because there is no risk of disrupting traffic with second protection switch.

19.3.2 The Protection Protocol

Bidirectional protection requires coordination between both ends so that both directions of transmission can be switched. The protocol for bidirectional protection switching is carried in the K1 and K2 bytes of APS cells (see Figure 19.3) in the APS connection.

In group protection, the APS connection is the chosen from the group of connections. In individual protection there is no choice. There is an APS connection for both the working entity and for the protection entity, each at the appropriate ATM layer.

The protection protocol simply indicates to the other end whether the working or the protection entity has been selected and the reason for the selection. The other end may select differently if it is aware of a higher priority condition, but this mismatch should be resolved because it also indicates its choice and the reason for it.

A persistent mismatch between the two ends causes a fault notification to be generated to the Operations Systems. Different codes for the selector

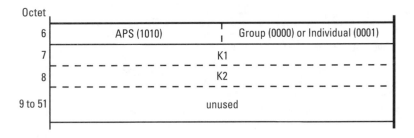

Figure 19.3 Function specific fields for APS.

status are used in 1+1 and 1:1 protection since this ensures that a notification is generated if the two ends try to use different types of protection.

19.3.3 Conditions, Commands, and States

Protection switching is influenced by signal conditions, by external commands, and by the internal status. The internal status has the least influence because it determines where the traffic is carried when no other considerations apply.

The highest priority conditions are not indicated in the protection protocol because they govern the priorities that are indicated in the protection protocol. A *Clear*, which clears all other conditions, has the highest priority. Next is a *Freeze*, which freezes the position of the local protection selection and prevents any change to the transmitted K1 and K2 bytes. The transmitted K1 and K2 bytes indicate all other priorities (see Figure 19.4). *Clear* and

Priority (K1)		Traffic (K2)	Description
LoP	(1111-0000)	W	Normal traffic locked or forced out of the protection entity
SF-P	(1110-0000)	W	Signal failure on the protection entity
FS-W	(1101-0001)	P	Normal traffic forced out of the working entity.
SF-W	(1011-0001)	P	Signal failure on the working entity.
SD-P	(1001-0000)	W	Signal degrades on the protection entity.
SD-W	(1000-0001)	P	Signal degrades on the working entity.
MS-P	(0110-0000)	W	Normal traffic manually switched out of the protection entity.
MS-W	(0101-0001)	P	Normal traffic manually switched out of the working entity.
WTR	(0011-0001)	P	Waiting to restore the working entity - revertive protection.
DNR	(0001-0001)	P	Do not restore the working entity - non-revertive protection.
NR	(0000-0000)	W or P	No requests active.

W: The use of the working entity for normal traffic is indicated by 0001-XXXX on the K2 byte for 1+1 protection and by 0000-XXXX for 1:1 protection.

P: The use of the protection entity for normal traffic is indicated by 0000-XXXX on the K2 byte for 1+1 protection and by 0001-XXXX for 1:1 protection.

Note: The four most significant bits of K1 indicated the priority and the least four indicate the affected transmission entity.

Figure 19.4 Indications Transmitted on the K1 and K2 Bytes

Freeze conditions must be removed locally because they are not communicated and there is no higher priority indication from the remote end.

A Lockout of Protection (LoP) is the highest priority indicated by the K1 and K2 bytes. This has a similar effect on the normal traffic as a failure on the protection entity and the normal traffic is forced back onto the working entity despite its condition. The next highest priority is the forced switching of the traffic out of the working entity, which has a similar effect to failure of the working entity.

Signal degrades have a lower priority that signal failures, but like signal failures a problem on the protection entity take priority over a problem on the working entity and forces traffic back onto the working entity. Manually forced switching has a lower priority than signal conditions, but like signal conditions the switching of traffic back onto the working entity takes priority.

The lowest priority state is the No Request (NR) state. In this state the location of the traffic is determined its past history. In revertive switching the Wait To Restore (WTR) state, which has traffic on the protection entity, takes immediate priority over the NR state and the WTR state will decay to the NR state with traffic on the working entity. In nonrevertive switching the Do Not Revert (DNR) state, which likewise has traffic on the protection entity, is the state that takes immediate priority over the NR state. Unlike the WTR state the DNR state is stable and the NR state cannot be reached from it directly.

19.4   Paths and Connections

Basic ATM call control has been extended to include VP connections [3], in addition to VC connections, and this has helped in the development of Soft Permanent Virtual Connections (S-PVCs)[4]. S-PVCs are automatically rerouted by call control signaling if a failure occurs. Unlike protection switching, the route used for the connection after a failure is not configured in advance.

Additional thought has also been given to multipoint connections, which have more than two end-points.

19.4.1   Switched Virtual Paths

Switched Virtual Paths (SVPs) are modeled [5] as an independent supplementary service (see Chapter 9 on ATM modeling). If the broadband customer profile contains this service (see Figure 19.5) then VP switching is enabled for the customer.

UNI accesses that support VP switching are enhanced to identify pools of VP level resources at the physical ATM interfaces. Each pool is controlled by the UNI signaling and represents a pool of bandwidth and VPI ranges at a transmission convergence. A pool is identified and labeled with its base VPCI (Virtual Path Connection Identifier) as indicated in Figure 19.5.

These UNI access cannot be configured for associated signaling, because associated signaling does not apply to switched VPs. A signaling VC cannot lie within the VP it controls because the VP connection is end-to-end while the signaling VC must terminate in the switch.

NNI accesses that support VP switching must also be enhanced to identify the pools of VP resources at the physical ATM interfaces (see Figure 19.6).

19.4.2 Soft Permanent Virtual Connections (S-PVCs)

Except for the absence of UNI signaling at the calling end, an S-PVC is established like a normal on-demand connection. Instead of using UNI signaling, the ATM switch at the calling end establishes the connection because its Operations System (OS) has configured it to do so (see Figure 19.7).

In the modeling of S-PVCs, a soft PVC object holds the directory number of the called party (see Figure 19.8). It also indicates the Virtual Path Connection Identifier (VPCI) and the Virtual Channel Identifier (VCI) at the destination if these are needed. The soft PVC object can also notify the Operations System (OS) of problems with the connection.

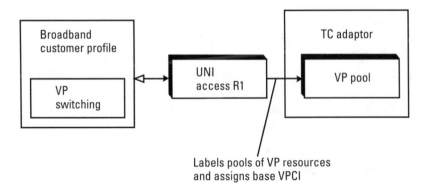

Figure 19.5 Modeling of switched VPs at UNIs.

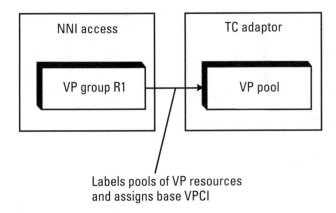

Labels pools of VP resources
and assigns base VPCI

Figure 19.6 Modeling of switched VPs at the NNI.

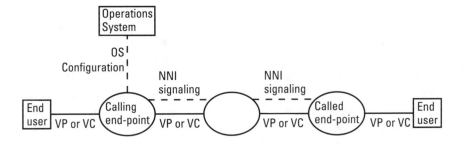

Figure 19.7 Soft PVC management and signaling.

The object can identify the directory number and UNI access of the calling party and hold details of its VPCI and VCI for the connection. It can also identify an ATM traffic descriptor for the connection and hold information about reconnection attempts.

19.4.3 Multipoint Connections

The simplest connection topology is a single connection between two points, known as a point-to-point connection. Point-to-point connections are normally bidirectional (or duplex) and if they are established by signaling then typically one of the ends is the calling end and the other end is the called end.

Multipoint connections, which have more than two end-points, have been traditionally described in terms of a single root and multiple branches.

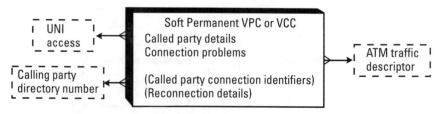

Note: Dotted lines indicated optional objects

Figure 19.8 Modeling of soft permanent virtual connections.

This allows the root to be identified with the calling end. This tradition is restrictive because it binds the call signaling to the connection topology. In particular, the root-and-branch terminology is misleading when third party signaling is used because the connection is not then controlled by one of its ends.

The root-and-branch terminology also encourages a simplistic view of directionality. The connection segments are normally assumed to have the same directionality. If they are unidirectional (or simplex) and from the root to the branches then the connection is described as point-to-multipoint. If they are unidirectional in the opposite direction, then the connection is described as multipoint-to-point. If they are bidirectional the term bidirectional point-to-multipoint is used.

Unfortunately, in a general connection topology interconnected ends are joined by connection segments, each of which can be bidirectional or unidirectional. If they are unidirectional then one of the ends is the source and the other end is the sink. The traditional approaches to multipoint connections do not handle the general case. Regardless of whether the connections are established by signaling or by management, the focus has been on the root-and-branch view.

There have also been problems with the management modeling because the approach used by the ATM Forum and by ITU-T [6]. was not consistent with the generic model developed in ITU-T [7] because the generic model did not take account of the early work on ATM. The ATM Forum subsequently has led the way both on the signaling for multipoint connections and on the management modeling. Unfortunately the root-and-branch paradigm has persisted.

The root-and-branch paradigm may be challenged in a future version of the ITU-T model, but the support of general multipoint connections by conventional call control signaling may take even longer to develop. It may be easier to support general multipoint connections through an IP application protocol.

19.5 Traffic, Services, and Quality

In the original ATM modeling, intermediate points of connections could include a description of the traffic, in particular its Peak and Sustainable Cell Rates (PCR and SCR), the associated Cell Delay Variation (CDV) tolerances, and the Maximum Burst Size (MBS). These parameters can have different values for an incoming (ingress) cell flow and for an outgoing (egress) flow. There are also different values for an OAM flow, but only for its PCRs and CDV tolerances.

When the ATM Forum revised its modeling, it introduced the idea that a traffic descriptor could be a profile that the intermediate points could refer to (see Figure 19.9). This view has since been accepted by the ITU-T.

Unfortunately, there are fundamental differences between the ITU-T and ATM Forum characterizations of ATM connections, and these differences are not resolved by the adoption of a common approach to the management modeling. The most obvious difference is the additional ITU-T Quality of Service (QoS) classes. The differences polarize the choice of equipment because they make interworking difficult, and they often undermine the adoption of the ITU-T standards.

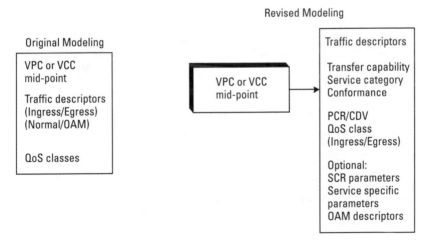

Figure 19.9 Modeling of traffic parameters and QoS.

19.5.1   ATMF Service Categories and ITU-T Transfer Capabilities

Aside from the QoS classes, the approaches of the ATM Forum and the ITU-T are approximately similar. The ATM Forum defines service categories that resemble ITU-T's ATM transfer capabilities (ATCs) [8]. For example, the ATM Forum's Constant Bit Rate (CBR) service category corresponds to the ITU-T's deterministic bit rate (DBR) transfer capability (see Figure 19.10). Likewise the variable bit rate (VBR) service category corresponds to the statistical bit rate (SBR) transfer capability.

The correspondence between the service categories and the transfer capabilities is not exact. In particular there is no service category that corresponds to the ATM block transfer (ABT) transfer capability, and there is no transfer capability that corresponds to the unspecified bit rate (UBR) service category.

There are also cases where the correspondence exists but is incomplete. Available Bit Rate (ABR) is fully specified by the ATF Forum, but only partially specified by ITU-T. The same is true for Guaranteed Frame Rate (GFR). Furthermore, the VBR service category can be either real-time or non real-time but only one of varieties of the corresponding SBR transfer capability (SBR.1) can be real time.

Real-time rate traffic requires tight constraints on the Cell Transfer Delay (CTD) and the Cell Delay Variation (CDV). These constraints apply regardless of whether the bit rate is constant or variable. For variable bit rate traffic, the Sustainable Cell Rate (SCR) is less than the Peak Cell Rate (PCR)

ATMF Service Category	ITU-T Transfer Capability	Notes
CBR	DBR.1, DBR.2	DBR.2 = DBR.1 + explicit OAM
rt-VBR, nrt-VBR	SBR.1, SBR.2, SBR.3	All SBR varieties can be non-real-time, but only SBR.1 can be real-time
UBR	✕	No ITU-T definition.
ABR	ABR	Partial ITU-T definition.
✕	ABT.1-IT, ABT.1-DT ABT.2-IT, ABT.2-DT	ABT.2 = ABT.1 + explicit OAM
GFR	GFR.1, GFR.2	Partial ITU-T definition.

Figure 19.10 ATMF service categories and ITU-T transfer capabilities.

and a burst of cells transmitted at the PCR must not exceed the Maximum Burst Size (MBS).

The UBR service is intended for non-real time applications, such as file transfer and email, which can be supported on a best-effort basis. Unlike other services, UBR does not have a specified CTD or Cell Loss Ratio (CLR). A PCR may be indicated for UBR, but it is not enforced. Congestion control is not included in UBR, but may be provided by an application.

Congestion control is included in ABR. Feedback on congestion is provided through Resource Management (RM) cells. Like UBR, there is no specified delay or delay variation for the cells. ABR does have a specified MCR (Minimum Cell Rate) and a PCR.

In addition to a guaranteed minimum throughput, GFR allows dynamic access to additional bandwidth. GFR has no flow control protocol to handle congestion. Instead, GFR discards the complete PDUs (Protocol Data Units) of AAL5 instead of discarding single cells. This is sensible because applications will need to send entire frames again if one of the cells has been lost. The GFR service has a Maximum Frame Size (MFS) in addition to an MCR, a PCR and an MBS.

19.5.2 Traffic Parameters, QoS Parameters, and QoS Classes

The ATM Forum approach to QoS also differs from the ITU-T approach. This is because the ITU-T approach [9] specifies QoS classes, whereas the ATM Forum approach specifies the individual QoS parameters. The ATM Forum approach also allows certain QoS parameters (peak-to-peak CDV, maximum CTD, and CLR) to be negotiated for individual connections.

ITU-T has enumerated five QoS classes. Two of these relate to constant and variable bit rate data, and two relate to connection-oriented and connectionless data. Each of these has a defined Cell Loss Ratio (CLR) that depends on the Cell Loss Priority (CLP) bit in the cell header. They also have an average CTD and a CDV that are independent of CLP since cells arrive in sequence regardless of their CLP value. The fifth and final ITU-T class has no specified QoS parameters because it is for best-effort traffic.

In the ATM Forum approach there are no QoS classes and the QoS parameters are directly related to the service categories (see Figure 19.11). This resembles the way that traffic parameters are also related to the service categories.

Service Category	QoS Parameter			Traffic Parameter					
	CLR (CLP=0,1)	CTD	CDV	PCR	CDVT	SCR	MBS	MCR	MFS
CBR	Yes	Yes	Yes	Yes	Yes	No	No	No	No
rt-VBR	Yes	Yes	Yes	Yes	Yes	Yes	Yes	No	No
nrt-VBR	Yes	No	No	Yes	Yes	Yes	Yes	No	No
ABR	Maybe	No	No	Yes	Yes	No	No	Yes	No
GFR	Maybe	No	No	Yes	Yes	Yes	Yes	Yes	Yes
UBR	No	No	No	Yes	Yes	No	No	No	No

Bursts cannot arrive faster than is consistent with the sustainable rate. This means the minium time between bursts is Ti:

Ti = Maximum Burst Size (MBS)/Sustainable Cell Rate (SCR)

Figure 19.11   Traffic parameters and QoS parameters.

19.6   Summary

The initial ATM OAM flows have been extended to include Automatic Protection Switching (APS) flows and to allow activation/deactivation flows for forward performance monitoring without backward reporting. The format of the location field initially defined for loopback and alarm reporting cells has been clarified. Fields that link performance reporting cells with forward monitoring cells, and that indicate the number of received cell blocks with severe errors have also been added.

The new APS cells support protection switching of individual connections and groups of connections, at both the Virtual Path (VP) and at the Virtual Channel (VC) layers. These cells support bidirectional 1+1 protection, where traffic is transmitted on both a working route and a protection route in both directions and the better route is selected at the destination. They also support bidirectional 1:1 protection, where the protection route may only carry extra traffic under normal conditions and this is pre-empted by the protected traffic if the normal route fails. Unidirectional 1+1 APS can also be used but this does not need OAM cells for coordination since the only decision is the choice of the correct flow at the destination.

ATM call control protocols have been extended to include VP connections. Switched virtual paths (SVPs) are modeled as a supplementary ATM service. SVPs are especially useful because they extent soft permanent virtual connections (S-PVCs) to the VP layer. S-PVCs are configured at one end only and then established across the network by call control signaling. They are re-established in the event of failure without the need for prior configuration of the protection route as in APS, although the outage time may longer.

Much of the traditional work on multipoint connections naively assumes that one of the end-points (the root) is special and the others (the branches) have similar properties. This remains the paradigm in most conventional call control, but it is being challenged in the management modeling. IP based control of general multipoint connections may be necessary if the delay of changing the traditional multipoint paradigm is to be avoided.

There are significant differences between the ATM Forum and ITU-T on the characterization of ATM connections, despite the adoption of a common approach on the management modeling. There is a reasonable degree of correspondence between the ATM Forum's service categories and ITU-T's transfer capabilities but this is not exact. Unfortunately the ITU-T has additional quality of service classes whereas the ATM Forum ties the applicability of parameters directly to the service categories. This polarizes the deployment of equipment, often to the advantage of the ATM Forum's concepts.

References

[1]   *B-ISDN Operations and Maintenance: Principles and Functions,* ITU-T Recommendation I.610, 1999.

[2]   *ATM Protection Switching,* ITU-T Recommendation I.630, 1999.

[3]   *Switched Virtual Path Capability,* ITU-T Recommendation Q.2766.1, 1998.

[4]   Soft PVC Capability, ITU-T Recommendation Q.2767.1, 1998.

[5]   Enhanced Broadband Switch Management, ITU-T Recommendation Q.824.7, 1999.

[6]   *Asynchronous Transfer Mode (ATM) Management of the Network Element View,* ITU-T Recommendation I.751, 1996.

[7]   *Generic Network Information Model,* ITU-T Recommendation M.3100, 1995.

[8]   *Traffic Control and Congestion Control in B-ISDN,* ITU-T Recommendation I.371, 1996.

[9]   *B-ISDN ATM Layer Cell Transfer Performance,* ITU-T Recommendation I.356, 1996.

20

Optical Technology for IP

It is a riddle, wrapped in a mystery, inside an enigma.
—Winston Churchill

IP datagrams are often wrapped in ATM, inside SONET/SDH. Winston Churchill's comment was made about Russia in 1939 and heralded the start of a war. The war concerning optical transmission of IP is about whether it is the ATM layer or the SONET/SDH layer that should be removed, or whether it should be both.

If IP datagrams are Russia, then WDM technology is Switzerland since it remains neutral.

20.1 Background

Much of the transmission used between Internet routers is not especially suitable because it was developed for conventional telecommunications. It has layers of electronic multiplexing that the Internet Protocol (IP) does not need. These layers add cost and complexity and contain overheads that reduce efficiency. They also partition the links, and this can interfere with the effective concentration of traffic.

The conventional approach to carrying IP in a broadband network is to:

1. Use ATM Adaptation Layer 5 to indicate the datagram boundaries on ATM cells;

2. Assign the ATM cells to a Virtual Channel (VC) within a Virtual Path (VP);

3. Transport the VP in a SONET/SDH Virtual Container;

4. Transmit the SONET/SDH frames over an optical carrier.

IP datagrams could be mapped directly into an optical bit stream using the widely accepted Internet Point-to-Point Protocol (PPP) [1]. There is no need for the complexity and constrained data rates of SONET/SDH, and ATM itself could also be eliminated.

New optical technology is also more suitable for the core of the broadband network and for Internet links than conventional technology because it is being developed more quickly. It is increasingly clear that existing SONET/SDH rates are too low, both for network core and for the high capacity links to Service Providers (SPs).

There have also been attempts to develop optical technology more generally for the Internet. Although it does not lend itself to the routing of individual datagrams, it has been applied to bursts of datagrams and to establish and release optical communication channels. Optical technology has also been used to create prototypes of purely optical Local Area Networks (LANs).

20.2   IP over Serial Data Links

For remote IP access, datagrams are sent to Internet Service Providers (ISPs) over a serial data link that has been designed for conventional telecommunications. This also occurs if IP datagrams are sent over directly over SONET/SDH. In both cases PPP can be used to indicate the datagram boundaries.

PPP relies on some method of recovering byte synchronization from the received data stream because it is a byte-oriented protocol, in contrast to the bit oriented High-level Data Link Control (HDLC) protocol [2]. Once a receiving end recovers byte synchronization it can proceed to recover frames and packets. Byte stuffing differentiates the frame boundaries from the contents of the frame. A particular byte, the framing byte, indicates a frame boundary (see Figure 20.1). Another byte, the escape byte, ensures that the data transmitted in a frame does not mimic a frame boundary. If a data byte

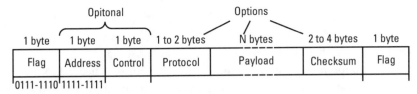

Figure 20.1 illustration labels:

Opitonal | Options

1 byte	1 byte	1 byte	1 to 2 bytes	N bytes	2 to 4 bytes	1 byte
Flag	Address	Control	Protocol	Payload	Checksum	Flag

0111-1110 1111-1111

Notes:
1) There is an option to omit both the address and control fields because neither are needed if the control field takes its default value (0000-0011).
The default value of the control field does not allow the receipt of numbered frames to be acknowledged because it indicates an unnumbered frame.

2) If the first bit of the protocol field is zero then the other bits specify a network layer protocol such as IP.
If the first bit of the protocol field is one of the other bits specify a negotiation protocol, such as LCP, which negotiates the PPP options, or a network protocol specific NCP. The NCP for IP allows IP addresses to be assigned.

Figure 20.1 PPP frame format.

matches the framing byte or the escape byte then the escape byte followed by a second byte is sent instead.

PPP both frames the payload and has a checksum that enables corrupted frames to be detected. It is sensible to detect and discard corrupted frames as soon as possible because this reduces the traffic load.

The length of the checksum is flexible and the address and control fields are optional. This flexibility allows the PPP overhead to be minimized. The value of the address field is fixed and indicates that no address matching is needed so the destination of the frame is any machine that receives it.

The protocol field enables PPP options to be negotiated. It identifies the nature of the payload, including payloads that contain PPP's Link Control Protocol (LCP). Four of the LCP packet types allow PPP options to be proposed and negotiated. In addition to the options on the length of the checksum and the elimination of the address and control fields, there are options on the length of both the payload and of the protocol field.

Options are negotiated during the establishment of the data link. If something is received but cannot be understood then an LCP rejection packet is returned. Other LCP packet types are used for echo testing and for debugging. At the end of a session, LCP packets that denote the termination of the link are sent.

The Protocol field can also indicate that the payload contains some other protocol. This could be an alternative to LCP, a network layer protocol such as IP, or a Network Control Protocol (NCP). NCPs are associated with specific network layer protocols. The NCP for IP allows an IP address to be assigned for the duration of a session. ISPs can find this useful because they have a finite number of Internet addresses at their disposal.

For nonpermanent accesses some form of security is needed to protect against infiltration. Passwords are the most common approach, but users often create passwords that are easy to guess. Callback systems increase the security by establishing a connection back to the proper user location, but infiltrators have been known to tap outgoing telephone lines and callback limits user mobility. Dialers, which require a changing authentication number to be entered, are an effective way of preventing infiltration especially when a large number of people require access.

20.3   Optical IP Transmission

PPP allows IP datagrams to be carried directly over serial links, avoiding the traditional telecommunications multiplexing hierarchy. IP can be carried directly over SONET/SDH without the need for an ATM layer. ATM can also be carried by cell-based transmission, avoiding the need for the SONET/SDH layer (see Figure 20.2). IP can also be carried directly over fiber using Wavelength Division Multiplexing (WDM).

20.3.1   IP over SONET/SDH

The Internet Engineering Task Force (IETF) has defined the method of carrying IP over SONET/SDH [2]. IP datagrams are encapsulated into PPP frames that are mapped into SONET/SDH virtual containers (see Figure 20.2).

The use of PPP at Gbps rates has been questioned because it uses byte stuffing. Byte stuffing allows a malicious user to attack the link by transmitting strings of flag bytes. This reduces the bandwidth available to other users since the link would have to transmit twice the number of bytes to avoid confusion with framing flags. It is not a problem if bit stuffing is used because bit stuffing at worst inserts a zero after a string of five ones.

Using the SONET/SDH virtual containers to carry IP is in contrast with ATM PONs (see Chapter 18). ATM PONs use SONET/SDH rates but define their own framing structure. IP datagrams could likewise be carried at SONET/SDH rates as an HDLC bit stream, but this could not really be termed IP over SONET/SDH since the SONET/SDH multiplexing hierarchy and overheads would be missing.

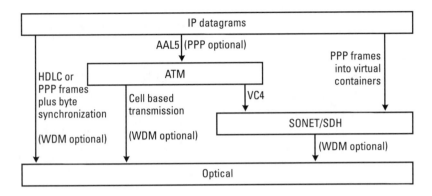

Figure 20.2 IP over different layers.

20.3.2 ATM Versus SONET/SDH

An alternative to eliminating the ATM layer is to eliminate the SONET/SDH layer. This has lead to a debate over the relative merits of the two layers, and a "plague on both your houses" view by some members of the IETF Internet Engineering Task Force (IETF).

The main criticism that is made of the ATM transport layer is that it includes an ATM "cell tax." Part of this cell tax is the overhead of five octets that are added to every 48-payload octets. There are also cells that are not completely filled and entire IP datagrams must be retransmitted if a single cell is lost. The cell tax is at least 10% because of the cell overhead and a figure of 25% has been suggested [4].

The ATM cell tax is all the more damaging because networks are often engineered so that the traffic on the links does not exceed 85% of their capacity. The combination of cell tax and traffic engineering could result in links that only carry about 60% of their nominal data rate.

Although the overhead for IP over SONET/SDH has been estimated at 2%, this is misleading because it does not take account of the complexity of the SONET/SDH multiplexing hierarchy and the granularity of SONET/SDH. Although ATM cells may not be filled to capacity, the same is true for SONET/SDH links. If there are several links between two points then on average half of a link will not be used because in some cases the last link will be almost full while in others it will be almost empty.

A more valid criticism of ATM is that increasingly higher speed ATM chip sets need to be developed. This is more of reason to avoid both layers since higher speed SONET/SDH chip sets would also be required. The criticism may also become less significant as time goes on since the number of wavelengths used in WDM transmission creates a limit to the data rates on a single channel.

The issue is not whether the SONET/SDH layers should be removed, but whether the ATM layers should remain. An advantage of WDM is that by having a number of independent optical channels on the same fiber, both ATM and IP can be carried directly. From the point of view of an ISP, it is sensible to avoid both the ATM and the SONET/SDH layers because this reduces the equipment cost.

20.3.3 IP over WDM

By allowing different wavelengths to carry different channels, WDM allows optical fibers to be used more effectively because it is difficult for the optoelectronic components in a single channel to operate fast enough to use the full capacity of the fiber.

Optical filters allow the 1550 nm window on an optical fiber to support up to 30 channels with 1nm between channels (see Figure 20.3). This corresponds to about 100 GHz per channel or a total optical channel bandwidth of about 3,000 GHz. WDM with tens of channels in this window is known as Dense WDM (DWDM) to differentiate it from early WDM with one channel at 1310 nm and another at 1550 nm.

Next generation DWDM can increase the number of channels to over a hundred with tens of GHz per channel. If coherent optical technologies [5] are used then this is called Optical Frequency Division Multiplexing (OFDM).

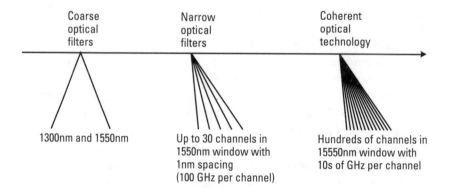

Figure 20.3 WDM evolution.

WDM resembles radio transmission more than it resembles traditional multiplexing because it uses the same physical medium at different frequencies and does not impose a format on the data for transmission. This is a significant advantage because it means that WDM technology can be used regardless of whether or not ATM and SONET/SDH are used.

20.4 Optical IP Networks

New routers with higher capacity are needed for the Internet because of the constantly increasing demand for Internet based services. This has led to considerable interest in the use of optical fiber technology.

Optical routing differs from the optical transmission of IP because it is concerned with the direction of packets to different destinations, rather than the transmission of packets to a remote router. A true optical packet switch would direct packets according to the destination address in the packet headers. This is difficult even if an optical switch could create paths quickly enough, because optical technology is not suited to the analysis of addresses.

Optical burst switching and optical flow switching have been devised to overcome these difficulties.

20.4.1 Optical Burst Switching

Optical switching is paradoxical. Although an optical switch can carry data at a high rate, it is slow at switching the data from one destination to another. Burst switching counters this by directing bursts of packets towards the same destination [6].

The destination of the burst is established by a control packet that an upstream router sends on a separate channel (see Figure 20.4). For simplicity and speed the control packet is not acknowledged and this means that the burst may fail.

Optical burst switching needs electronic routers to groom the IP traffic into bursts and generate the control packets. The bursts and control packets are sent to the optical routers on WDM channels over optical fibers. Although optical routers are not effective alone, their high throughput when combined with electronic grooming routers make them suitable for the optical backbone of an internet.

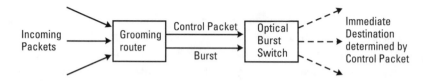

Figure 20.4 Optical burst switching.

20.4.2 Optical Flow Switching

A more effective way to introduce optical switching may be to create temporary optical paths across a network because this is better suited to its high throughput and slow switching speeds. These long duration, high data rate flows are created and controlled by an optical flow protocol [6]. The simplest optical flow switches use a single wavelength on a WDM transmission since this avoids wavelength conversion.

Optical flow switching is similar to shortcut routing over an ATM network (see Chapter 21) and the temporary optical paths are similar to ATM connections. Optical flow switching is simpler because there is no need for ATM adaptation or ATM call control.

As in burst switching, electronic routers are needed to groom the traffic so that the optical technology can be used effectively (see Figure 20.5). Flow switching differs from burst switching because the optical flows are established for some time and not just for a burst of packets. Greater judgment is required when setting up an optical connection than when setting up a

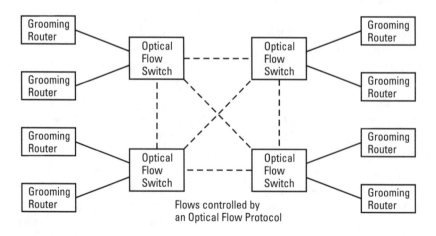

Figure 20.5 Optical flow switching.

virtual ATM connection. This is because greater resources are being committed, especially if there is no wavelength conversion as the wavelength cannot then be re-used within the optical medium.

20.4.3 WDM LANs and WANs

It is easier to design an optical LAN than it is to design an optical router because a LAN can use wavelength selection instead of optical switching. Since the attenuation of optical fiber is so low, an optical LAN could be spread over a large geographical region. If such a LAN linked the gateway routers of conventional LANs, then it would be equivalent to a WAN.

Each node of the WDM LAN is linked to the center of the LAN by two fibers [6]. One fiber is used for transmission and the other for reception. At the center of the WDM LAN there is a star coupler that connects all of the incoming signals to all of the outgoing fibers (see Figure 20.6). An optical wavelength is assigned to each of the nodes. The nodes communicate by transmitting on the different optical wavelengths.

At the nodes, both the optical transmitters and the optical receivers should be tunable. One should be tunable to so that a wavelength can be assigned to each node without manual installation. The other must be tunable to allow communication with the other nodes. Some form of control channel is needed to coordinate the use of the optical wavelengths, and a Media Access Control (MAC) protocol may be needed both for this channel and for the data channels.

Tuning must be fast to prevent nodes spending more time tuning than communicating. Since each wavelength can carry data at Gbps and there may be a few thousand bits in a small packet, tuning should take less than a microsecond. The speed and cost of the optical transceivers are the main drawbacks to the implementation of WDM LANs.

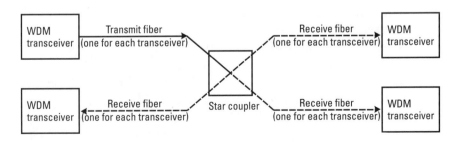

Figure 20.6 WDM LAN.

20.4.4 Superimposed Optical Topologies

In a traditional optical transmission system, the physical topology of the optical fiber determines the optical topology because the physical medium only carries a single optical flow. WDM allows the creation of a separate optical topology because it allows different optical signals to follow different routes across the physical medium [6].

Frequency sensitive switches can direct different optical wavelengths along different paths. Even more sophisticated optical topologies are possible if the optical switches change the wavelength of the optical signal because then the same wavelength can carry different signals in different parts of the optical network.

20.5 Summary

Conventional telecommunications links have multiple layers that add cost, complexity and inefficiency to Internet communications. Direct optical transmission allows these layers to be eliminated. Optical technology can also be applied to routers and LANs.

The Point-to-Point Protocol (PPP) can carry IP datagrams across serial links regardless of the transmission technology. PPP contains options that can be negotiated through its Link Control Protocol (LCP) and it supports Network Control Protocols (NCPs) that are specific to the network layer protocol, such as the NCP that allows dynamic allocation of IP addresses.

PPP can carry IP datagrams directly over SONET/SDH links within the SONET/SDH multiplexing hierarchy. Alternatively the SONET/SDH layer can be eliminated and IP can be carried in ATM cells with cell-based transmission. The main criticism that is made of the ATM layer is that it adds an ATM "cell tax" that reduces the useable data rate. There is also a significant inefficiency if SDH links are unfilled, and the real issue is whether both layers should be eliminated.

Wavelength division multiplexing (WDM) is an effective way of using optical fibers because it is easier to use the fiber to capacity with several optical channels since optoelectronic components are too slow to do this with a single channel. Although WDM allows IP to be carried directly, it also allows more traditional transmission to be used on the same optical fiber.

Optical routing of individual datagrams is not feasible because, although optical switches can carry high data rate streams, optical technology is not appropriate for address analysis and the optical switching of data streams to different destinations is slow. Optical burst switches direct bursts of datagrams towards a destination that is determined by a control packet. Optical flow switches create temporary optical paths across an optical network under the control of an optical flow protocol. Optical routers are more appropriate to backbone applications because they need electronic routers to groom the traffic and control the bursts or flows.

A local area network (LAN) can be completely optical since it can substitute wavelength selection for optical switching. A WDM LAN can cover the same geographical area as a conventional WAN, but it requires low cost optical components that can be tuned to different wavelengths within microseconds.

WDM technology also allows different optical topologies to be superimposed on the same optical fiber technologies.

References

[1]   Simpson, W., ed., *The Point-to-Point Protocol (PPP)*,
       RFC 1661, (http://www.ietf.org/rfc/rfc1661), July 1994.

[2]   Tanenbaum, A. S., *Computer Networks*, New Jersey: Prentice-Hall, 3rd ed., 1996.

[3]   Malis, A., and W. Simpson, *PPP over SONET/SDH*,
       RFC 2615, (http://www.ietf.org/rfc/rfc2615), June 1999.

[4]   Manchester, J., et al., "IP over SONET," *IEEE Communications Magazine*, Vol. 36, No. 5, May 1998.

[5]   Gillespie, A., *Access Networks*, Norwood, MA: Artech House, 1997.

[6]   Modiano, E., "Topics in Lightwave WDM-Based Packet Networks," *IEEE Communications Magazine*, Vol. 37, No. 3, March 1999.

21

The Way Forward

I do benefits for all religions—I'd hate to blow the hereafter on
a technicality.

—Bob Hope

One way to find out what the future holds is to visit it. A number of time machines have now been designed. They operate like escalators. You can travel up them into the future and then back down to when they were first built. Time paradoxes, such as those involving killing your own grandfather when he was a baby, seem to be resolved by quantum physics. It appears possible to build a temporal escalator by correctly aligning several fast-spinning neutron stars. To a theoretical physicist the difficulty involved is merely a technicality.

A short trip up a temporal escalator might reveal a future where all communications are carried on the Internet. Those who believe that this is the hereafter sometimes seem to express their views with a religious intensity. A slightly more heretical view is that the future will also permit other networks, such as ATM, but that they will be managed and controlled using the Internet Protocol. Only infidels, surely, could believe that other protocols are the true way forward.

A more enlightened view may be that no single approach has a monopoly on the truth. Carrying all communications on the Internet, like building a temporal escalator, appears to be quite possible. The problems involved are mere technicalities.

21.1   Background

The historical approach of routing IP datagrams on a best-effort basis is not adequate for the increasing numbers of Internet applications that require a guaranteed Quality of Service (QoS). It has also been recognized that datagrams need to be routed more quickly and that better mechanisms to control congestion are required. These pressures have led to the development of Multiprotocol Label Switching (MPLS) [1]. MPLS has similarities to ATM, and these similarities have raised the question of whether MPLS can replace ATM.

MPLS is supported by flow labeling in version six of the Internet Protocol (IPv6) [2], but it was concern about the shortage of IP addresses that drove the development of IPv6. IPv6 also supports secure communications and real-time IP applications. The Resource Reservation Protocol (RSVP) [3], which is an IP signaling application, was developed to allow the resources needed to ensure a particular QoS to be reserved for a labeled flow of datagrams. The QoS parameters for a labeled flow of Internet datagrams are similar to those for an ATM connection.

The classical approach to IP over ATM is not efficient because it does not allow machines to communicate over an ATM connection unless they belong to the same IP network or subnetwork. The Next Hop Resolution Protocol (NHRP) [4] was developed to overcome this imposed constraint by allowing ATM shortcuts between machines on different IP networks. Unfortunately, the shortcuts supported by NHRP are limited by the IP network topology and do not take account of the required QoS. As an alternative to coordinating NHRP for shortcuts with RSVP for QoS, a revised form of RSVP has been proposed.

Attempts have also been made to eliminate the costs and complexity of having control and routing protocols in both the IP domain and the ATM domain. There are aspects of the VB5 standards that may assist this simplification.

21.2   Multiprotocol Label Switching

Multiprotocol Label Switching (MPLS) may revolutionize the Internet because it supports different qualities of service through changing the way that IP datagrams are routed [5]. The traditional Internet approach is that each datagram is routed independently and delivered on a best-effort basis. MPLS allows datagrams to be sent along a specific route that is identified by a label. MPLS is not specific to a particular protocol. In addition to IP, it can support proprietary protocols such as Novell's Internetwork Packet Exchange (IPX).

MPLS enables routers to operate faster because they can route packets according to their label without consulting a routing table. Labels can be present in the frame layer or the packet layer, or a labeling header can be added to a packet. Label Switch Routers (LSRs) examine these short labels instead of the long destination addresses within packets.

MPLS can make networks more efficient by increasing the utilization of the links [6]. Increased link utilization is achieved through better congestion control, since MPLS allows labeled streams to be redirected as the traffic on the link becomes too high. Before MPLS, IP traffic could only be redirected by changing the routing metric, which is too crude as it affects all of the traffic on the link.

The Internet is resilient to failure because it finds alternative routes automatically when a link fails. MPLS can use this resilience to create an alternative path when a labeled path fails. The time to restore a path in this way is similar to that required to update routing tables when a link fails. For real-time traffic it would be better to set up two paths and have the destination chose between them because it creating a new path would take too long.

21.2.1 Forwarding Equivalent Classes

Routers perform the routing and forwarding functions that are needed when Local Area Networks (LANs) are interconnected to form Wide Area Networks (WANs). Routing is the process of deciding the next hop from the routing table and a packet's network-layer address. Forwarding is the subsequent step of determining the appropriate interface and sending the packet along it.

Routers maintain their routing tables by communicating with their peers using protocols such as the Border Gateway Protocol (BGP) and Open Shortest Path First (OSPF). A router can use routing information to partition its forwarding Internet address space into similar regions called Forwarding Equivalency Classes (FECs) that are assigned labels.

The router can attach the appropriate FEC label to the packets that it forwards. Subsequent routers can route the packets more quickly because they can determine the FEC and the route to it without having to re-examine the network-layer address.

21.2.2 Creating Label-Switched Paths (LSPs)

Switching according to labels is similar to ATM, but in MPLS entire packets are switched instead of ATM cells. An LSR can change the label of a packet (see Figure 21.1) in the same way that an ATM switch can change a cell's Virtual Path (VP) and Virtual Channel (VC) identifiers. A complete Label Switched Path (LSP) is similar to an ATM connection.

FEC labels can be recursively stacked so there can be many MPLS layers, unlike ATM where there are only the VP and VC layers. The streams that LSPs carry can also be merged since an LSR can give the same outgoing label to packets with different incoming labels (see Figure 21.2). In theory, ATM can merge cell flows but in practice most implementations do not support this.

LSRs can use the Label Distribution Protocol (LDP) to distribute labels to their peers. LSRs only distribute labels to their upstream peers, but they may do so spontaneously or in response to a request from an upstream peer. If the distribution is independent then each LSR allocates labels for the

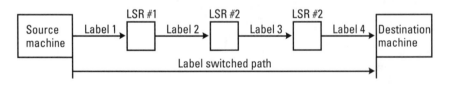

Figure 21.1 Label switched path (LSP) in MPLS.

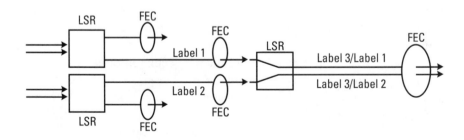

Figure 21.2 Recursive stacking of labels for merged MPLS flows.

streams that it recognizes. If the distribution is ordered the label allocation is propagated upstream from the egress LSR, which has no downstream peer.

Loops in LSPs can be detected by including a path vector field, which identifies the LSRs along the LSP, in the distribution message. If an LSR finds its own identifier in the path vector field then it rejects the message. Otherwise the LSR appends its own identifier and passes the message upstream.

Loops can be prevented if an upstream check of the new path vector field is performed when an LSP is changed. The check is propagated upstream until the ingress LSR is reached. The ingress LSR can then confirm that the new LSP is valid. Any upstream LSR that finds its identifier in the path vector causes the check to fail.

21.2.3 MPLS Versus ATM

The similarity between MPLS and ATM raises the question of ATM is really needed. ATM is an attempt to create a universal communications fabric, but it was not devised with IP in mind. The overwhelming acceptance of IP suggests that it may be simpler to use IP with MPLS as the universal communications fabric.

In the classical IP overlay model, ATM is a secondary technology that allows some of the shortcomings of IP to be circumvented. Like Frame Relay, it provides virtual circuits between IP routers and allows the superposition of a virtual topology on the physical network topology. It also supports traffic management because utilization statistics can be gathered for the virtual circuits, and the network can be adjusted to improve its operation.

If MPLS alone was to be used then no separate ATM network would be needed, but investment in MPLS would still be required and there would also be the problem of existing ATM services that MPLS could not immediately support. A reduction in the number of hops on routes through the elimination of ATM nodes should improve reliability, but using ATM short cuts between routers has a similar effect.

There is a fallacy that ATM does not scale well when it used for IP. It is incorrect to claim that the number of ATM virtual circuits must grow as the square of the number of nodes, because ATM was designed to prevent this. Even if there were no switched virtual circuits, for broadband access the number of circuits only grows as the number of end-users since there are no direct connections between end-users.

It has also been suggested that it is difficult to perform ATM adaptation at high data rates, but this implicitly assumes that switching IP datagrams is easier. The weakness of this and other points in favor of MPLS does not prove that it is better to use ATM as a secondary technology, but it does indicate that it is difficult to dismiss ATM.

21.3   Internet Protocol Version 6

The Internet gained global eminence in the 1990s, based on version four of the Internet Protocol (IPv4) [7]. When IPv4 was devised, its 32 bit addresses were thought to be more than adequate. Because of the rapid growth of the global Internet and the convenient but inefficient two-level structure of IPv4 addresses, they are now regarded as too small. This led to the development of version six of the Internet Protocol (IPv6) [2]. Version five of the Internet Protocol was bypassed due to a typographical error that is now entrenched.

There was also pressure to support real-time audio and video communications, and to make communication more secure. Real-time communication requires resources to be allocated to a communications channel in a much the same way as for connection-oriented communications. Secure communication requires authentication of the sender and encryption of the data.

21.3.1   Datagram Format

IPv6, like IPv4, supports best-effort delivery of datagrams, but the datagrams have a completely different format. The revised format contains larger addresses and supports new features, and it has been rationalized to ease the processing of datagrams.

The 32-bit IPv4 address range has been extended to 128 bits, but the number of header fields in the datagram has been reduced (see Figure 21.3). Unlike IPv4, an IPv6 datagram begins with a base header that is always 40 octets long. In IPv6, optional information is included in extension headers, not in a single variable length header. This modularity simplifies the addition of options and it helps routers to process datagrams faster because they often need only examine the fixed length base header [8].

21.3.2   Fields and Headers

The first field in both types of datagram is the same, because this is the *Version* field that lets routers differentiate between them. This is six for IPv6. The last two fields in an IPv6 base header are the Source address and

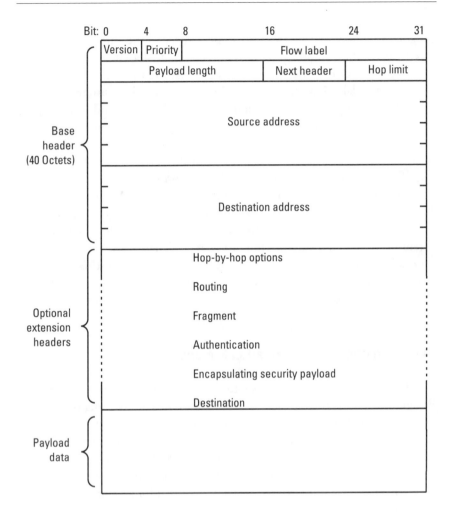

Figure 21.3 IPv6 datagram format.

Destination address, which are the last two fields in an IPv4 header that contains no optional fields or padding. In IPv6 these are four times larger.

The base header and the extension headers contain a Next header field that indicates the type of header that follows. Extension headers also indicate their length in their Header extension length field. The base header does not indicate its length because this is fixed, and instead indicates the start of the payload through its Payload length field.

The inclusion of the Flow label field is the most significant IPv6 innovation because it supports guaranteed QoS. It allows datagrams to be assigned to a flow that has resources reserved across the Internet. Each flow has defined values of, and variations in, delay and data rate. A datagram can be routed according its flow label because all datagrams from the same origin with the same label have the same destination and routing options.

Routing options are contained in the Routing and Hop-by-hop options extension headers. The order in which extension headers follow the base header is fixed. The base header also contains the Priority field, which sets the relative priority of datagrams from the same source, and the Hop limit field, which is equivalent to the IPv4 time-to-live field.

Security and authentication functions use the Authentication and Encapsulating security payload extension headers. Fragmentation information, which is included in the IPv4 header, is included in the Fragmentation extension header. To simplify the processing of the datagrams, information on destination routing that is processed only by the final destination is contained in a different Destination extension header from that which is processed at all destinations specified for optional hop-by-hop routing.

21.4   The Resource Reservation Protocol

The Internet has traditionally operated on the basis of best-effort delivery of datagrams. Traditionally, each datagram is routed independently with no guarantee of delivery and with no account taken of the data rate, error rate, or delay sensitivity of applications. This is not adequate for real-time applications, and the Resource Reservation Protocol (RSVP) [3] has been developed to overcome this. RSVP is an IP signaling application that allows resources to be reserved along a route so that the QoS required for an application can be guaranteed.

RSVP messages can either be carried raw within IP datagrams or encapsulated within UDP (User Datagram Protocol). The requests (Path messages) include the IP address of the original sender and the QoS for the sender's intended traffic. The QoS information resembles the parameters that are used in ATM connections. If IPv4 is used then the sender includes a user port, but if IPv6 is used then the sender may include the flow label instead.

Each intermediate RSVP capable node updates the Path message with its address. If the Path message is intended to inform the destination of the capabilities of the intermediate nodes, then the node also adds information about its resources (see Figure 21.4). The final destination can then trigger the reservation of resources by sending a Resv message back upstream.

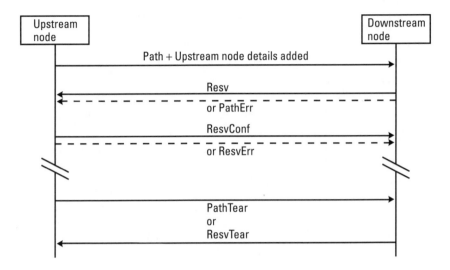

Figure 21.4 RSVP resource reservation and release.

If there is a problem with the received Path message then the intermediate node sends a PathErr message back upstream. Likewise if there is a problem with the received Resv message then a ResvErr message is sent back downstream. The ResvConf message provides positive confirmation of the Resv message.

If the sender does not periodically transmit additional Path messages then the path will timeout and the reserved resources will be released. When a session is over, the sender should send a PathTear message downstream to release the reserved resources immediately. Other nodes that detect the loss of a path should also send a PathTear message downstream or a ResvTear message upstream.

21.5 Shortcut Routing

Traditionally, IP routers were designed for flexibility and ATM switches were designed for speed. ATM switches can operate faster because they use hardware that switches fixed-format cells according to an exact match with the identifiers in their headers. IP routers are more flexible because they use software that routes variable length datagrams according to the best match with the destination in their headers.

IP may not be the universal communications solution, but it is the ubiquitous solution. In the past, ATM has offered faster and better quality communications, although this may not remain true in the future. Unfortunately the conventional use of LAN Emulation (LANE) and Logical IP Subnets (LISs) has discarded the benefits of ATM so that ATM can support IP in a simple and naive way.

Traditionally, datagrams are routed hop-by-hop from source to destination, regardless of any other communications infrastructure that could be used. Shortcut routing enables routers to take advantage of an underlying communications infrastructure (see Figure 21.5). Although PSTN and Frame Relay networks can also support shortcut routing, ATM is of greater interest because ATM connections can provide more bandwidth and wider QoS.

To avoid creating a large number of brief ATM connections, discretion is needed when creating shortcuts. Shortcuts may be unnecessary in many, if not most cases, especially when no QoS is specified. However, those shortcuts that are created are likely to have a significant effect on the traffic because they should carry large numbers of datagrams.

21.5.1 Policy-Based Shortcuts

Normally two nodes would cooperate to establish an ATM shortcut. A simpler alternative is for a single end to establish an ATM shortcut based on a

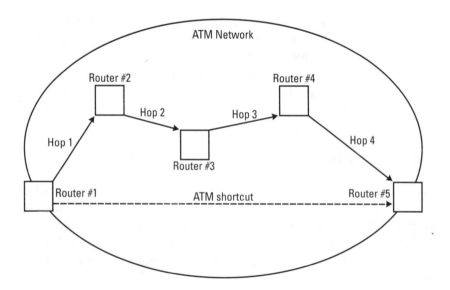

Figure 21.5 Shortcut ATM routing.

specified policy. Single nodes can be configured to create shortcuts for certain IP addresses according to rules that determine when a shortcut is appropriate. Shortcuts are more appropriate for IP datagrams carrying TCP (Transmission Control Protocol) payloads than to datagrams carrying UDP (User Datagram Protocol) payloads because TCP involves sessions of transactions whereas UDP can carry an independent datagram.

Policy based shortcuts are a precursor to more sophisticated methods because these too should not create large numbers of unnecessary shortcuts or shortcuts that are inappropriate. However, policies for shortcuts can take account of QoS and can support multicast communication. A policy-based approach may also use unconventional means of creating shortcuts.

21.5.2 Next Hop Resolution Protocol

ATM shortcuts that depend on the IP topology can be identified using the Next Hop Resolution Protocol (NHRP) [4]. NHRP was devised to regain some of the benefit of ATM by allowing routers in different LISs to communicate directly using shortcuts.

If the IP route to a destination lies within an ATM network then NHRP allows the destination's ATM address to be determined from its IP address. If the IP route leaves the ATM network then NHRP allows the ATM address of the router closest to the destination to be determined. Once the ATM address is known, an ATM shortcut can be set up.

Unlike a normal Address Resolution Protocol (ARP), NHRP operates across different networks and does not rely on broadcasting. Machines using NHRP to create shortcuts are called Next Hop Clients (NHCs). NHCs register with Next Hop Servers (NHSs) using NHRP so that other NHCs can obtain their ATM addresses. Caches of addresses in clients and servers can be configured or built up over time.

The only NHRP message that is not a request or a reply is the error indication. There are requests and replies for address registration, address resolution and address purging.

A client that wishes to resolve an IP address sends a request to its server (see Figure 21.6). Like ARP messages, NHRP resolution requests contain the IP and ATM addresses of the originating client so that a response can be returned and an ATM shortcut opened. When a server receives a request concerning one of its clients it returns the corresponding ATM address.

When a server receives a request that is not for one of its clients, it forwards it downstream towards the included IP address if it can do so without sending it outside the ATM network. Otherwise it responds with its own

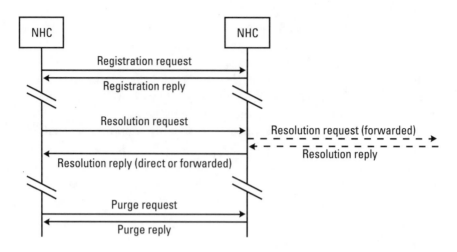

Figure 21.6 NHRP transactions.

ATM address since this is the longest ATM shortcut that it can guarantee. There is no need for servers to exchange address information because forwarding requests is sufficient. Servers along the normal IP route must forward requests instead of generating their own requests because only one shortcut to the original client is needed, not a shortcut to each of the servers.

If no server is capable of guaranteeing a shortcut route then a negative reply is returned. If the route leaves the ATM network via a bridge then an aggregate reply, which maps several IP addresses to a single ATM address, may be returned. Upstream servers may use replies to update their caches and then reply directly to requests. Using replies to extend caches can result in information being out of date, so it is sensible to purge old entries. Requests can also specify an authoritative response that can only be answered by the true server.

The NHRP is useful because it removes the artificial limit on communication between routers on the same ATM network. Unfortunately the messages of NHRP do not carry QoS information and they do not directly support multicast communication. They also require routers to be able to handle both IP communication and ATM call control.

21.5.3 Topology-Independent Shortcuts

ATM shortcuts may also be identified if ATM signaling carries information about the location of routers. This is effective because it allows the best shortcut to be identified regardless of the topology of the IP network. PNNI

Augmented Routing (PAR) [9] supports this approach. Unfortunately, like NHRP, the approach requires the co-ordinated use of both Internet protocols and ATM signaling and routing protocols.

It is cheaper and simpler to use a single type of protocol for shortcuts. This can be achieved if the ATM network is controlled by an IP protocols, instead of conventional ATM protocols. ATM routing can be determined by IP routing protocols, since PNNI (Private Network-Network Interface) already uses a routing protocol similar to OSPF (Open Shortest Path First). The creation and release of ATM connections can also be controlled by IP protocols. Different shortcuts can be created for different types of traffic, and policies can be used to determine when shortcuts are needed.

Discarding conventional ATM protocols is a drastic step. It may be the sensible way to avoid the disadvantages of using both IP protocols and ATM protocols since IP is well established and the deployment of ATM protocols has lagged behind the deployment of ATM networks.

21.6 RSVP for ATM

The Resource Reservation Protocol (RSVP) provides a means of reserving resources for a labeled flow of IP datagrams. ATM virtual connections have a specified Quality of Service (QoS). There is a similarity between the QoS parameters of RSVP and those for ATM. RSVP and ATM should be able to collaborate to provide the QoS needed for labeled flows of IP datagrams.

21.6.1 Quality in IP and ATM

The traditional ways of using ATM for IP pay no attention to QoS. They view ATM as another network technology that can be used for best-effort delivery of datagrams. This does not do justice to ATM because it ignores the guaranteed QoS that ATM can provide. It also does not do justice to IP because it ignores IP's need for QoS.

For RSVP and ATM to collaborate, a mapping is needed between the two sets of QoS parameters. The relationship between ATM connections and IP flows also needs to be clarified since a single ATM connection can support several IP flows. ATM connections can also be switched or permanent, and this affects the address translation between the two domains.

NHRP [4], which allows an IP address to be resolved into the ATM address of a shortcut, is no great help here. This is because NHRP does not scale well, since it is limited to administrative domains, and because NHRP does not carry QoS information. It should also be possible to have pre-established ATM shortcuts between routers since this can make the process faster.

21.6.2 Making Shortcuts with Reservations

An alternative approach is to modify RSVP to carry ATM addresses [10]. This is sensible because it does not require new co-ordinated processes for RSVP and NHRP that handle QoS. RSVP *Path* and *Resv* messages can be simply extended to include the ATM addresses of machines that support ATM shortcuts.

The ingress machine at the start of possible shortcut, inserts its ATM address into the Path message to indicate that it is prepared to use a shortcut (see Figure 21.7). This shortcut must have ATM QoS parameters that correspond to the IP QoS parameters in the Path message. This ATM address is carried transparently by the intermediate routers that are connected to the ATM network. If the route never leaves the ATM network, the destination machine returns a Resv message with its own ATM address so that an ATM shortcut with the desired QoS can be used.

If the route leaves the ATM network then the egress router extracts and records the ATM address of the ingress router when it forwards the Path message. When the egress router receives a returned Resv message, it inserts its ATM address. The intermediate routers carry this transparently back to the ingress router. The ingress router extracts the address and the two can

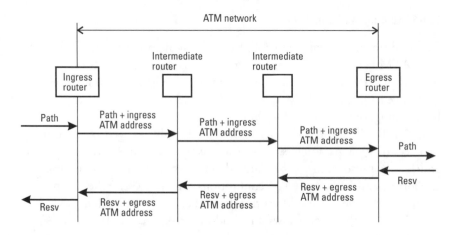

Figure 21.7 Single domain shortcut using modified RSVP.

then use an ATM shortcut because they know each other's ATM addresses and the required QoS.7.

This simple approach is limited by the topology of the IP network. A more sophisticated modification to RSVP that allows several ATM addresses to be added allows this limitation to be overcome.

21.6.3 Proxy Addresses and VB5

Modifying RSVP to carry ATM addresses is not ideal if it assumes a new conventional switched ATM connection will be set up between machines. It is better if existing connections can be re-used and if connections do not need to rely on conventional ATM call control.

Addresses for permanent ATM connections are an additional challenge because the identifiers used to specify their end-points are not globally unique, but local to an ATM interface or transmission convergence. The VB5 architecture can help here because it allows local ATM identifiers to be concatenated with identifiers for user ports and VB5 interfaces to provide a unique address. The VB5 architecture also allows connections to be set up on demand by some other party using proxy signaling.

In proxy signaling the creation of connections is triggered by a mechanism other than conventional user signaling. Proxy signaling is useful when interworking with narrowband networks since the narrowband signaling can trigger the creation of an appropriate ATM connection.

An alternative trigger can be the receipt of QoS requirements and ATM addresses from modified RSVP messages. In the VB5 architecture these can trigger the use of the B-BCC (Broadband Bearer Channel Connection) protocol to create connections. The VB5 standards also permit connections to be created using the RPC (Remote Procedure Call) protocol over IP across an X-VB5 interface.

21.7 Summary

Multiprotocol Label Switching (MPLS) is revolutionizing the operation of the Internet by allowing datagrams to be assigned to flows with specified Quality of Service (QoS). Label Switch Routers (LSRs) route datagrams according to their labels, giving faster routing and better control of congestion. A Label Distribution Protocol (LDP) distributes labels and allows routing loops to be identified and prevented.

Label Switched Paths (LSPs) are the MPLS equivalent of ATM virtual connections. Like ATM, MPLS can support traffic management and direct links between routers, but unlike ATM MPLS allows labels to be stacked indefinitely and it is an integral part of the Internet and is not a separate network.

Version six of the Internet Protocol (IPv6) supports MPLS through the flow label in the IPv6 datagram header. IPv6 was developed because of a shortage of Internet addresses and to support secure communications and real-time applications. The format of IPv6 datagrams has been rationalized to make routing easier. Options are included in extension headers that follow the fixed length base header in a defined order, and the fields in the base header have been simplified.

The Resource Reservation Protocol (RSVP) allows resources that support a desired QoS to be allocated to flow of datagrams. Path messages, which include QoS information, are updated with information about each RSVP router and forwarded towards the destination. The destination can send a Resv message to reserve the identified resources. Paths timeout and their resources are released if they are not refreshed with new Path messages.

QoS is also needed on shortcut ATM connections between routers that are conventionally separated by a number of hops. Although extending conventional ATM signaling allows the best IP shortcuts to be identified, this has the cost and complexity of co-ordinating both IP and ATM protocols. A more efficient approach is to use IP applications instead of conventional ATM control protocols.

Policies are needed to prevent a shortcut being created each time a single UDP (User Datagram Protocol) datagram is sent. Shortcuts that are determined by policies alone can be flexible since they can take account of QoS and permanent connections.

Suboptimal ATM shortcuts that depend on the IP topology can be identified at the IP layer using the Next Hop Resolution Protocol (NHRP). The NHRP allows Next Hop Clients (NHCs) to ask Next Hop Servers (NHSs) to resolve an IP address into an ATM address. Clients register with servers that either answer queries directly or pass them towards the IP destination.

NHRP does not take account of the relationship between IP flows and the different types of ATM connections, or of how ATM QoS should be used for IP services. A better approach is to map IP QoS onto ATM QoS and to include ATM addresses into RSVP messages. ATM addresses can be added and removed as the IP route crosses different ATM networks, or they can be stacked to allow the best shortcut to be identified regardless of the IP topology.

The addresses used for shortcuts should allow a variety of ATM connections to be used, including those established by unconventional signaling and proxy signaling. Unique global addresses for permanent connections can be defined using the VB5 standards and these also support connection control through IP application protocols.

References

[1]   Awduche, D., et al., *Requirements for Traffic Engineering Over MPLS*, RFC 2702, (http://www.ietf.org/rfc/rfc2702.txt), September 1999.

[2]   Deering, S., and R. Hinden, *Internet Protocol, Version 6 (IPv6) Specification*, RFC 2460, (http://www.ietf.org/rfc/rfc2460.txt), December 1998..

[3]   Braden, R., ed., Zhang, L., Berson, S., Herzog, S., Jamin, S., *Resource ReSerVation Protocol (RSVP)—Version 1 Functional Specification*, RFC 2205, (http://www.ietf.org/rfc/rfc2205.txt), September 1997

[4]   Luciani, J., et al., *NBMA Next Hop Resolution Protocol (NHRP)*, RFC 2332, (http://www.ietf.org/rfc/rfc2332.txt), April 1998

[5]   Viswanathan, A., et al., "Evolution of Multiprotocol Label Switching," *IEEE Communications Magazine*, Vol. 36, No. 5, May 1998.

[6]   Li, T., "MPLS and the Evolving Internet Architecture," *IEEE Communications Magazine*, Vol. 37, No. 12, December 1999.

[7]   Postel, J., *Internet Protocol*, RFC 791, (http://www.ietf.org/rfc/rfc791.txt), September 1981.

[8]   Stallings, W., "IPv6: The New Internet Protocol," *IEEE Communications Magazine*, Vol. 34, No. 7, July 1996.

[9]   Dumortier, P., "Toward a New IP over ATM Routing Paradigm," *IEEE Communications Magazine*, Vol. 36, No.1, January 1998

[10] Cocca, R., and S. Salsano, "Internet Integrated Service over ATM: A Solution for Shortcut QoS Virtual Channels," *IEEE Communications Magazine*, Vol. 37, No. 12, December 1999.

Acronyms and Abbreviations

A/D Activation/deactivation

AAL ATM adaptation layer

ABR Available bit rate

ABT ATM block transfer

ABT-DT ATM block transfer with delayed transmission

ABT-IT ATM block transfer with immediate transmission

ACM Address complete message

ADSL Asymmetrical digital subscriber line

AESA ATM end-system address

AFI Authority and Format identifier

AIS Alarm indication signal

AL   Alignment

AN   Access network

ANM   Answer message

ANSI   American national standards institute

AOC   ADSL overhead control channel

APS   Automatic protection switching

ARP   Address resolution protocol

ARPA   Advanced Research Projects Agency

ARPAnet   Advanced Research Projects Agency Network

ATC   ATM transfer capability

ATM   Asynchronous transfer mode

ATMARP   ATM address resolution protocol

ATMF   ATM Forum

ATU   ADSL transceiver unit

ATU-C   ADSL termination unit-central

ATU-R   ADSL termination unit-remote

ASN.1   Abstract syntax notation one

AU   Administrative unit

AUG   Administrative unit group

BASize   Buffer allocation size

B-BCC   Broadband bearer channel connection

BCD   Binary coded decimal

BER   Bit error rate (in transmission context)
basic encoding rules (in ASN.1 context)

BGP   Border gateway protocol

B-ICI   Broadband inter-carrier interface

BIP   Bit interleaved parity

B-ISDN   Broadband ISDN

B-ISUP   B-ISDN user part

BOM   Beginning of message

BOOTP   Bootstrap protocol

Btag   Beginning tag

CAP   Carrierless amplitude and phase

CBR   Constant bit rate

CC   Continuity check

CD   Compact disk

CDV   Cell delay variation

CDVT   Cell delay variation tolerance

CER   Cell error ratio

CES   Circuit emulation service

CIDR   Classless interdomain routing

CIP   Carrier identification parameter

CLIP   Calling line identification presentation

CLIR   Calling line identification restriction

CLP   Cell loss priority

CLR   Cell loss ratio

CMIP   Common management information protocol

CMR   Cell misinsertion ratio

COLP   Connected line identification presentation

COLR   Connected line identification restriction

COM   Continuation of message

CPCS   Common part of the convergence sublayer

CPI   Common part indicator

CRC   Cyclic redundancy checksum

CS   Convergence sublayer (in ATM adaptation context)
Capability set (in ATM services context)

CS-1   Capability set 1

CS-2   Capability set 2

CS-3   Capability set 3

CSI   Convergence sublayer indication

CSP   Carrier selection parameter

CTD   Cell transfer delay

CTP   Connection termination point

CUG   Closed user group

DBR   Deterministic bit rate

DCHP   Dynamic host configuration protocol

DDI   Direct dialing in

DMT   Discrete multitone

DNR   Do not revert

DNS   Domain name system

DSL   Digital subscriber line

DSLAM   Digital subscriber line access multiplexer

DSP   Digital signal processing (in electronic context)
Domai- specific part (in ATM address context)

DSS2   Digital subscriber signalling system No 2

DWDM   Dense wavelength division multiplexing

EFCI   Explicit forward congestion indication

EGP   Exterior gateway protocol

EOC   Embedded operations channel

EOM   End of message

Etag   Ending tag

ETSI   European telecommunications standards institute

FDDI   Fiber distributed data interface

FEC   Forward error correction (in coding context)
Forwarding equivalency class (in IP context)

FERF   Far end receiver fail

FEXT   Far end cross talk

FM   Fault management

FSAN   Full services access network

FTP   File transfer protocol

FTTB/C   Fiber to the building/curb

FTTCab   Fiber to the cabinet

FTTH   Fiber to the home

Gbps   Gigabits per second

GFC   Generic flow control

HDLC   High-level data link control

HDSL   High-speed digital subscriber line

HEC   Header error control

HFC   Hybrid fiber coax

HTTP   Hypertext transfer protocol

IAA   IAM acknowledge

IAM   Initial address message

IAR   IAM reject

ICMP   Internet control message protocol

IETF   Internet engineering task force

IGMP   Internet group management protocol

IGP   Interior gateway protocol

IHL   Internet header length

IP   Internet protocol

IPv4   Internet protocol version 4

IPv6   Internet protocol version 6

IPX   Internetwork packet exchange

ISDN   Integrated services digital network

ISP   Internet service provider

ITU   International telecommunications union

ITU-T   ITU-Telecommunications standardisation sector

GCAC   Generic call admission control

GFR   Guaranteed frame rate

GGP   Gateway-to-gateway protocol

GHz   Gigahertz

ISO   International organization for standardization

Kbps   Kilobits per second

kHz   Kilohertz

KISS   Keep it simple, stupid

LAN   Local area network

LANE   LAN emulation

LCP   Link control protocol

LDP   Label distribution protocol

LI   Length indicator

LIS   Logical IP subnet

LLC   Logical link control

LoP   Lockout of protection

LSP   Label switched path (in MPLS context)
Logical service port (in VB5 context)

LSR   Label switch router

LUP   Logical user port

MAC   Media access control

Mbps   Megabits per second

MBS   Maximum burst size

MCR   Minimum cell rate

MCSN   Monitoring cell sequence number

MFS   Maximum frame size

MIB   Management information base

MID   Multiplex identification

MPLS   Multiprotocol label switching

ms   millisecond

MSN   Multiple subscriber number

MTP-3   Message transfer part level 3

NCP   Network control protocol

NE   Network element

NEXT   Near-end cross talk

NHC   Next hop client

NHRP   Next hop resolution protocol

NHS   Next hop server

N-ISDN   Narrowband ISDN

N-ISUP   Narrowband ISDN user part

nm   Nanometer

NNI   Network-network interface / network-node interface

NPC   Network parameter control

NR   No request

NSAP   Network services access point

NSF   National science foundation

NSFnet   National science foundation network

NT   Network termination

NVT   Network virtual terminal

OAM   Operations, administration, and maintenance

ODN   Optical distribution network

OFDM   Optical frequency division multiplexing

OID   Object identifier

OLT   Optical line termination

ONT   Optical network termination

ONU   Optical network unit

OS   Operations system

OSPF   Open shortest path first

OSS   Operational support systems

PABX   Private automatic branch exchange

PAR   PNNI augmented routing

PCR   Peak cell rate

PDH   Plesiochronous digital hierarchy

PDU   Protocol data unit

PLOAM   Physical layer operations, administration, and maintenance

PM   Performance monitoring

PNNI   Private network-network interface

PON   Passive optical network

PPP   Point-to-point protocol

PSP   Physical service port

PSTN   Public switched telephony network

PT   Payload type

PTSP   PNNI topology state packet

PUP   Physical user port

PVC   Permanent virtual connection

P-VCC   Permanent virtual channel connection

P-VPC   Permanent virtual path connection

QAM   Quadrature amplitude modulation

QoS   Quality of service

RARP   Reverse address resolution protocol

RAS   Remote access service

RDI   Remote defect indication

REL   Release

RFC   Request for comment

RIP   Routing information protocol

RLC   Release complete

ROM   Read-only memory

RPC   Remote procedure call

RM   Resource management

RSVP   Resource reservation protocol

RTMC   Real-time management coordination

SAAL   Signaling ATM adaptation layer

SAR   Segmentation and reassembly

SBR   Statistical bit rate

SCR   Sustainable cell rate

SDH   Synchronous digital hierarchy

SDT   Structured data transfer

SECBR   Severely errored cell block ratio

SID   Signaling identifier

SLIP   Serial line internet protocol

SM   System management

SMDS   Switched multimegabit data service

SMTP   Simple mail transfer protocol

SN   Sequence number (in ATM adaptation context)
Service node (in VB5 context)

SNAP   SubNetwork attachment point

SNMP   Simple network management protocol

SNP   Sequence number protection

SONET   Synchronous optical network

SP   Service provider

S-PVC Soft permanent virtual connection

SRTS Synchronous residual time stamp

SS7 Signaling system no. 7

SSCF Service-specific coordination function

SSCOP Service-specific connection-oriented protocol

SSCS Service-specific convergence sublayer

SSM Single segment message

ST Segment type

STM Synchronous transport module

STS Synchronous transport signal

STS-1 Synchronous transport signal level one

STS-N Synchronous transport signal level n

SVC Switched virtual connection

S-VCC Switched virtual channel connection

S-VPC Switched virtual path connection

TC Transmission convergence

TCP Transmission control protocol

TMN Telecommunications management network

TTP Trail termination point

TU Tributary unit

TUG   Tributary unit group

UBR   Unspecified bit rate

UDP   User datagram protocol

UNI   User-network interface

UPC   Usage parameter control

USB   Universal serial bus

VBR   Variable bit rate

VC   Virtual channel (in ATM context)
Virtual container (in SONET/SDH context)

VCC   Virtual channel connection

VCI   Virtual channel identifier

VCL   Virtual channel link

VCG   Virtual channel group

VDSL   Very-high-speed digital subscriber line

VoD   Video-on-demand

VP   Virtual path

VPC   Virtual path connection

VPCI   Virtual path connection identifier

VPI   Virtual path identifier

VPL   Virtual path link

VPG   Virtual path group

VPN Virtual private network

WAN Wide area network

WDM Wavelength division multiplexing

WTR Wait to restore